Amy Johnson

'Enigma in the Sky'

Amy Johnson

'Enigma in the Sky'

An Official Biography
by
David Luff

Airlife
England

Copyright © 2002 David Luff

First published in the UK in 2002
by Airlife Publishing Ltd

British Library Cataloguing-in-Publication Data
A catalogue record for this book
is available from the British Library

ISBN 1 84037 319 9

Typeset by Rowland Phototypesetting Ltd, Bury St Edmunds, Suffolk
Printed in England by MPG Books Ltd, Bodmin, Cornwall.
Distibuted in North America by
STACKPOLE BOOKS
5067 Ritter Road, Mechanicsburg, PA 17055
www.stackpolebooks.com

For a complete list of all Airlife titles please contact:

Airlife Publishing Ltd
101 Longden Road, Shrewsbury, SY3 9EB, England
E-mail: sales@airlifebooks.com
Website: www.airlifebooks.com

Satur Dierum
A person may die young, and yet be full of years

CONTENTS

INTRODUCTION

t was the morning of 17 December 1903 and a cold northerly wind was kicking the grains of sand off the dunes, as a few lonely seagulls squawked and circled under a leaden sky over Kill Devil Hill. Lying prone in the *Kitty Hawk Flyer* as it rocked gently to and fro in the gusting wind, the peak-capped pilot, Wilbur Wright, released the wire and the machine rushed down the take-off rail to stagger into the air. With the adrenalin tap on full flow, the innovative genius from Ohio saw the 100- and 200-foot markers slide slowly beneath the machine's wings to exceed the distance covered during the three previous flights of that morning. As the excitement mounted, the fragile biplane flew on to cover a distance of 852 feet at a height of no more than 15 feet for just under a minute.

Man had at last conquered the skies by making a controlled and sustained flight in a heavier-than-air machine for the first time. Meanwhile, 3000 miles away across the Atlantic on the north-east coast of England, a tiny, six-month-old baby girl was calling for the attention of her young mother. That cry has echoed down the corridors of time, and even today it still bids us listen.

When the wheels of Amy Johnson's travel-stained Gipsy Moth bumped down on the tussocky paddock of what passed for an airfield at Darwin in 1930, a legend was born. That a woman could enter into what was essentially a male-dominated profession and complete a solo flight in an open-cockpit biplane over a distance of 11,000 miles in nineteen days amazed the world. The fact that she was not a natural-born pilot and quite inexperienced at that, only added to the admiration that people felt for the bravery she had shown in traversing deserts, jungles and shark-infested seas without the help of radio or sophisticated navigational aids, equipment that would be considered essential today.

Amy Johnson's life could not have been more strange, if it had been scripted. This unknown slip of a girl, a solicitor's secretary from Kingston upon Hull, was to become an international star overnight, an icon, and then a folk-heroine. She epitomised courage and determination. Songs were written about her and people from all walks of life hummed, whistled or sang 'Amy, Wonderful Amy', the catch-tune of the day. Infant girls were named after her, and then schools, roads and even a rose. More importantly, the public at large took her to their hearts in a way that has seldom been repeated. She had the ability and charm to convey the common touch. Her 'girl next door' image and the self-deprecating way in which she could hold an audience endeared her to the nation. When she returned from Australia and flew into Croydon airport, it is estimated that somewhere in the region of a million people lined the streets to welcome her home as she was driven through the suburbs to London's West End. However, it was not the ordinary man and woman in the street alone who treated her with the respect they usually reserved for royalty, for the rich and famous vied equally for her attention.

Invariably, motivation is the key factor in achieving any kind of success. There has to be a strong inner drive, a powerful mainspring of action. There are grounds for suggesting that the rejection and humiliation Amy Johnson suffered from the man she adored and worshipped for six years provided the catalyst for the strength and determination she found within herself to succeed. The betrayal was no less painful for being a gradual one. However, it eventually brought about her metamorphosis from a woman emotionally traumatised into one who was to emerge with an iron will to succeed, if not in an affair of the heart then in some other way.

No one was more surprised by the fame and adulation which she received than Amy Johnson herself, and as a consequence she was forced to live in the limelight of a totally unexpected triumph. Of course she had expected to succeed, but not to that remarkable extent. At first it was a welcome novelty but then it became a case of slow suffocation. The demands made upon her to attend this dinner or this function, to open this garden party, to crown this Rose Queen, to present this trophy, cup or prize, became unbearable. Many who have fame thrust upon them today, particularly in the world of entertainment, seek refuge in a haze of alcohol or drugs. In Amy's case escape came through a series of collapses into various clinics, where she was treated for nervous exhaustion and depression.

Amy's paternal grandfather came to this country as a youth from Denmark to settle in Hull, where he quickly mastered the language and became the owner of a fishing smack. He subsequently took British nationality and changed his name from Anders Jorgensen to Andrew Johnson, and eventually founded a successful business as an importer and exporter of fish. One can only surmise that there may have been some causal link between Amy's susceptibility to emotional fragility and the Danish psyche – maybe a strand of melancholia lingered in the Nordic bloodline. The suicide of her younger sister, Irene, in 1929 at the age of twenty-four, and someone to whom she was particularly close, may also have been a factor. After Amy's death, the Johnson family was to face further suffering of Kennedyesque proportions when one of their sons-in-law was killed in action, and the youngest of their four daughters was subsequently to take her own life as a result.

Amy's life was very much entwined in this developing Greek tragedy, as she grappled with the perplexing paradox that one can be universally known and yet experience a devastating loneliness. She was drawn into a milieu where she was lionised by the rich and famous from all walks of life. It was one which covered a wide spectrum, ranging from royalty to presidents; from politicians and literary giants such as George Bernard Shaw, to celebrities such as Charlie Chaplin from the world of entertainment. However, in the midst of it all, she came to the realisation that it was the legend they really wanted and not *her*, the real person. She was once quoted as saying, 'I met everyone, went everywhere, and was miserable.' Escape came by seeking adventure in further record-breaking flights, partly to prove that the Australia flight was no mere fluke, but also to get away from what she perceived to be an artificial way of life.

Whatever Amy Johnson did, she did with an absolute intensity and compulsion. Whether she was out to break a long-distance flying record, or taking up some new sport such as horse-riding, skiing, gliding, competitive motoring or sailing, it mattered little. She was a compulsive letter writer and those letters which survive confirm her as a romantic, someone who longed for the reciprocity of love in marriage and yet was never able to find it. When James Allan Mollison came into her life in 1932, she believed that this aspect of her life would at last be fulfilled.

Much has been written and said of Jim Mollison and opinions on his character vary widely. Sir Alan Cobham described him as 'the most charming man I ever met', and fellow Scot and transatlantic pioneer pilot, John Grierson, thought him to be 'the most courageous and accomplished long-distance flier of his era'. Others, somewhat uncharitably, saw him as 'a libidinous drunk' and 'the compleat sociopath'. Maybe, in a way, he was all of these things combined, but when he proposed to Amy in Quaglino's restaurant she was swept completely off her feet. Together they became a *tour de force* without rival in the history of the Golden Age of aviation. Their frenzied ticker-tape welcome along Broadway in 1933, after they made the very first flight from the UK to the USA, a mammoth thirty-nine-hour non-stop flight, established 'the Flying Mollisons' as an international couple. Even the astronauts returning from their moon landing in the late sixties were not given the rapturous welcome that these early pioneers of aviation received.

Amy's marriage to Jim Mollison did little to enhance her emotional stability, since their time together was anything but a harmonious one. His heavy drinking, trail of debts and womanising led to a separation after only three years, whilst his 'kiss and tell' memoir, *Playboy of the Air*, only infuriated Amy and heightened the acrimony between them. There was more than one attempt at a reconciliation, but none of them worked. From then, Amy's considerable wealth drained away as the days of the aerial record-chasers waned, and she was forced to sell precious gifts and jewellery that she had acquired from her many admirers. It was a slow and gentle slide back into relative poverty.

Since I began researching for this book I have repeatedly been asked, 'Why another one on Amy Johnson? Surely, it has has all been done before.' Surprisingly, there have been relatively few books on the life of this amazing woman, and even then only one other (written over thirty years ago) which could in any way be described as a biography. Since that time, the vast number of letters that Amy wrote to Hans Arregger between 1922 and 1928 were auctioned and released into the public domain, so that now, for the first time, an unbiased view can be taken of them. Reading these letters today is much like drawing aside a curtain on a window of time, where one can not only observe the interaction between two people in a tortuous relationship, but also look into that particular period of history.

During the past four years I have sought to walk in my subject's footsteps, to be as it were her *doppelgänger*, a task that has required a certain amount of literary cross-dressing. Apart from the freedom of being able to draw fresh and important conclusions from her letters, I have also been fortunate in being able to interview or contact many new and hitherto unknown witnesses, people who were involved with events on the day of Amy's tragic death. The result has been that the traditionally accepted view that she baled out of her aircraft when its fuel was exhausted is seriously challenged. In addition, new material has come to light and the old, some of which was overlooked by previous researchers, has been sifted and examined carefully. Being the official biographer of her late ex-husband has helped, since I have also been able to present a balanced view of the part which he played in her life.

In the final chapter of the book I have attempted to assess Amy Johnson's historical position within the pantheon of early pioneering airwomen, a place which she shares with such acknowledged and illustrious names as Amelia Earhart, Anne Morrow Lindbergh, Beryl Markham and Jean Batten. It has been said that a biographer should be a novelist under oath and in this respect I have tried to live up to the aphorism. It is for the reader to judge whether I have succeeded or not. What is beyond dispute is that the name of Amy Johnson and her exploits will long be remembered. When asked by a colleague, just before she died, how she saw her fate as a wartime ferry pilot with the Air Transport Auxiliary, she replied: 'Oh, I'll probably end up in the drink and then there will be just a few lines in the newspapers – then they forget you!' How wrong can one be?

Amy Johnson CBE

Chapter One

CONVOY EAST 21

The first faint streaks of light were just appearing over the English Channel on the morning of Sunday, 5 January 1941, as Lieutenant Commander Walter Fletcher went up onto the bridge of HMS *Haslemere* and scanned the horizon through his binoculars. Thankfully, it appeared that Kesselring's young pilots and ground crews of *Luftflotte* II's squadrons were on stand-down and enjoying a quiet weekend. Either that or the low scudding cloud was putting them off.

The *Haslemere*'s captain had good reason to be apprehensive, for only six weeks earlier his ship had been under attack, not from enemy aircraft this time, but from the German long-range coastal batteries at Cap Gris-Nez. Over a four-hour period the convoy he was escorting was pounded by at least 170 rounds. One shell burst a few feet under the *Haslemere*'s port bow, wounding his first lieutenant and killing his sub-lieutenant beside him on the bridge, whilst the ship's side above the water line was pierced by shell splinters, damaging its superstructure, upper deck cabins and ventilators.

The 756-ton balloon barrage vessel *Haslemere*, formerly a fast cargo ship for Southern Railway, was now just one of the many escort ships shepherding seventeen merchant vessels, mainly colliers, towards their destination at Southend pier. They were accompanied by two Hunt-class destroyers, the *Fernie* and the *Berkeley*, a submarine chaser, four minesweepers, four other barrage balloon ships, four motor launches and two anti-submarine trawlers. With a protection ratio of one to one, it had become a costly business keeping southern Britain's power stations supplied with coal in wartime.

As the *Haslemere* made its way along the grey chalk outline of the Kent coast, its captain could just make out the spectre of a partially sunk collier. Its stern and bridge were 10 feet out of the water, its mast at a forlorn angle, and its bows completely submerged. It was just one of the many wrecks that dotted the coastline from Southampton to Southend, all victims of undetected mines or the enemy dive-bombers which had played such havoc with the convoys in the summer of 1940. Once they were hit, the skippers of these coasters would do their best to steer the ships towards the shore, so that they ran aground rather than sink completely. This at least gave the crews some chance to escape, and also left a possibility of the ship being salvaged at a later date. These ghostlike vessels were a grim reminder of the dangers that faced the brave men who kept Britain's lifeline open during these early days of the war.

The most dangerous period for any coastal convoy sailing through the Channel in those days was when they were passing through the Straits of Dover, for it was then that they came under the scrutiny of the Freya radar station on the cliff-tops at Wissant, almost opposite them in occupied France. At this point the balloons being flown by the escort vessels would be close-hauled down to 200 feet, in order to make them less visible on the German radar screens. Barrage balloon vessels such as the *Haslemere* carried an RAF crew of four, members of 952 Thames Barrage Squadron. It was their job was to operate the powered winch at the rear of the ship, which was used to raise or lower the balloon according to the height of the cloud base. Once a convoy was detected, it invariably came under threat either from the gun batteries, or an aerial attack. Less frequently, but just as deadly, was the possibility of being intercepted by an enemy torpedo boat.

On this particular morning, battle-hardened Walter Fletcher visibly relaxed as convoy CE 21 passed the North Foreland and formed into a single line ahead as it entered the mouth of the Thames Estuary. He knew that they were now out of the range of the enemy's coastal guns, but there was still the chance of a lone Junkers 88 slipping past the scrutiny of our own radar to attack out of low cloud. However, by far the greater hazard was caused by the enemy mine-laying aircraft based in the Frisian Islands, at Borkum and Sylt, intruders which operated by night to drop their parachute-borne mines.

These particular weapons of war were of two main types, the contact and the influence mine. One of the latter was the magnetic mine, which was detonated by the magnetic field of a passing ship; another was the acoustic mine, which reacted to the noise of a ship's engine vibrations. It was for this reason that the lead ship in the convoy would be a minesweeper, trailing its paravanes in the hope of cutting the mooring cables to which the mines were attached. The degaussing of ships, a method used to offset their magnetic influence, caused havoc with a ship's compass; whilst the throttling back of a ship's engine only went some way to protect it from an acoustic mine.

Although the mile-wide Channel routes along which the coastal convoys sailed were marked by red conical buoys, and were swept twice daily, one could never be certain that they were clean. German mine-laying aircraft had been especially active on the nights of 12 and 13 December 1940, when over fifty magnetic mines were dropped between Southend and the Isle of Sheppey. Each one was fitted with a four-day delayed-action mechanism, and such was their effectiveness that seven merchant ships were sunk within twenty-four hours. The tanker *Arinia* was one such victim when she disappeared suddenly in a gigantic ball of fire just off the Southend pier, with the loss of all but two of her sixty-nine crew members. By the end of that year over 6000 merchant seamen had given their lives. It was a death toll which the officers and crew of convoy CE 21 were very much aware of as they entered the final stage of their journey.

By early afternoon the convoy was sailing into the narrow channel of the Barrow Deep, heading for the Medway Gate and looking for the Port of London Authority pilot ship, which would be there to meet them. Heading the column, and closely followed by the destroyer HMS *Fernie*, was the minesweeper HMS *Pingouin*, flying one of the nine balloons to give aerial protection.

Cloud base was down to between 800 and 1000 feet on that day, a factor which did not prevent the gunners on the ships from keeping a sharp look-out for marauding aircraft. Their orders from the Admiralty, much to the chagrin of the Royal Air Force, were to fire on any unidentified aircraft approaching within 1500 yards. Unfortunately, aircraft recognition did not figure highly in the order of priorities of the Merchant Navy gunners, men who had received the most rudimentary training in the use of their First World War Lewis guns. After Dunkirk, it was very much a case of 'shoot first and ask questions later'.

The whole convoy now stretched line astern for almost 3 miles, each ship being required to keep station, no easy task when one considered the varying speeds and turning capabilities of so diverse a group of vessels. The colliers were now light in the water, for they had already disgorged their essential supplies of coal and coke to the wharves at Southampton and Portsmouth. These small ships, affectionately known as 'the coal scuttle brigade', were an essential part of the war effort, since the south coast alone was dependent upon 40,000 tons of coal a week to supply domestic and industrial needs.

With the convoy's speed now reduced to some 4 knots, it made its way slowly towards what would be the welcome sight of the pier at Southend. For some of these men it would mean the end of their regular six-day stint and a brief shore leave to be with loved ones. For other crew members, it meant joining a new convoy which would escort them and their 'Geordie' crews back up along the east coast to their ports of origin, such as Grimsby, Hull, Hartlepool and other bases along the coast to Tyneside.

Walter Fletcher probably looked forward to a spot of leave, when he would take the train to London and on to the family home in Princes Risborough. Sadly, that was not to be; before nightfall, he was to sacrifice his life in an heroic attempt to save the life of another.

Twenty-five-year-old Tom Mitchell had been conscripted into the Army early in 1940, and told to report to the barracks of the Kent Regiment at North Bottom, near Gravesend. It was a time when a great emphasis was being placed upon the strengthening of the anti-aircraft defences around Britain's ports, and in particular those in the area of the Thames estuary.

As a result of this policy the former Kent gardener was one of many who, soon after their mandatory square-bashing and a period of elementary training in gunnery, were posted to various units of the Royal Artillery. In his case, he was sent to be part of the 58th Heavy AA Regiment stationed on the Chetney Marshes, near Iwade, and not far from the port of Sheerness. It was here that he became one of the twenty or so gunners manning the four 4.5-inch guns of Battery 207 under the command of Captain Wilmore.

On this particular Sunday afternoon in January 1941 the young gunner was on a twenty-four-hour standby, which meant that he had plenty of time to write letters, to his girlfriend Thelma, or to his sister Rosemary. Not long before dusk the alarm was sounded for action stations. An unidentified aircraft, which was not responding to the unit's signals to give colours of the day, was somewhere overhead. As soon as Tom Mitchell took up his position alongside the gun-layer, readings were taken from height-finder and predictor, and they were ordered to open fire. The four guns each fired four rounds and then fell silent. If they had hit anything then they were completely unaware that they had done so.

Little more was said about the incident during that evening.

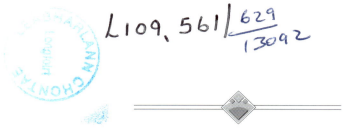

£109,561/629
13092

Attractive and vivacious nineteen-year-old Muriel Hollington, whose friends always referred to her as 'the Duchess' because she bore a striking resemblance to King Edward VII's paramour, was married to police officer Sam Hollington early in 1938. War clouds were on the horizon and they decided to delay starting a family until things improved. The couple rented a small, white, pebble-dashed house at the end of a cul-de-sac, which comprised some seventy dwellings in Ulster Avenue, Shoeburyness.

On a clear day the young housewife could look out across Shoebury Common, a grassed area which sloped gently down to the shoreline, and see Sheerness on the Kent side of the estuary. From her kitchen window there was always plenty of shipping to be seen as it plied its way to and from the capital, for this was probably the world's busiest waterway. Certainly the Port of London Authority was Europe's largest port, handling as it did over a third of all of Britain's trade.

Unfortunately, the avenue in which the Hollingtons lived was not an entirely peaceful one, since the house lay on the fringe of what has been known since Napoleonic times as the Artillery Range. From the edge of the Common, the Maplin Sands stretches for some 10 miles along the Essex coastline to the tip of Foulness Island. Along this strip the War Office restricted access to the civilian population and artillery practice shots were fired. On occasions the Hollingtons and their neighbours were warned to open their windows, particularly when the big 14-inch naval guns were to be fired. As a consequence of this military activity, civilian and RAF aircraft, especially those based at nearby Rochford, were permanently forbidden to fly within this area of the coast. Throughout the war it continued to be a 'no fly zone' to Allied aircraft.

Not long after war was declared, scaffolding was erected along Shoeburyness promenade as far as Southend Pier, shutting off the beach and its gaily painted beach huts to the public. This area subsequently became a militarised zone which was designated as HMS *Westcliff*, and it was along this part of the estuary that the wartime coastal convoys were assembled, briefed and dispersed.

During this time the locals saw the installation of the protective boom which stretched across the 6 miles of the estuary from Shoeburyness to Sheerness. The boom hung like a curtain net of steel to the sea-bed, and was designed to prevent enemy submarines from penetrating into the spacious mustering area of the convoys. It had only two narrow entrances for Allied shipping, the Yantlet Gate on the Essex side and the Medway Gate on the Kent side of the estuary. Similar booms were being constructed at Harwich and on the Humber.

As the war progressed, the *Luftwaffe*'s night-bombing raids increased along the south-east coast, and it was not long before several bombs fell quite near to the young couple's home. They were evidently aimed at the searchlight and light anti-aircraft unit of a coastal artillery regiment which was positioned uncomfortably close on the shoreline, in an area known as 'Gunners Park'. Not surprisingly, their presence was of concern to local residents, since the probing beam of its searchlight often became an irresistible target for the itchy fingers of some of the enemy's air-crews, as they made their way to and from London. After a while all the families in Ulster Avenue, apart from those of three railwaymen and one of Muriel's relatives, evacuated to safer areas. For Muriel and Sam Hollington there was no such escape, for his police duties meant that they had to stay put until the end of hostilities.

Muriel's most vivid and lasting memory of the war came on the afternoon of Sunday, 5 January 1941, not long after she kissed her husband goodbye as he set out for the late shift with the Southend Constabulary. He was office-based at that time due to a poor medical record, a factor which had prevented him from being conscripted for military service in 1939. The young housewife busied herself with the washing up, tidied round and then sat down in the lounge by the fireside to do some needlework.

She remembers it being a cold, miserable afternoon with a sea-mist. Intermittent rain and flurries of snow had been falling throughout the morning. As daylight began to fail on that wintry afternoon, she found it difficult to see the stitches she was trying to put into place. Her first instinct was to switch on the table-lamp beside her, but then she remembered the strict blackout rules that were in force. It was time to go outside into the garden to the back of the house and close the wooden shutters.

Pulling a jacket over her shoulders she stepped outside, and just as she slammed the shutters closed, she heard the deafening roar of an approaching aircraft. Almost sixty years later, she still recalls that moment with absolute clarity.

> It was a twin-engined aeroplane which seemed to be almost at the height of the chimney-pot of the house. It swept out into the estuary and disappeared from my view into the murk. Its engines were making a peculiar noise. I feel sure that if it had passed further down beyond the end of the cul-de-sac, it would have hit the elm trees by the old people's home. I immediately sensed that it was not an enemy aeroplane, for there

had been no air-raid warning. I just *knew* it was one of ours, and that it was in trouble. This being a no-fly zone we never *ever* saw low-flying aircraft over our houses, it was so unusual. It came in from the Rochford area. Whether it was searching for the aerodrome there, I'm not sure.

The first remark that Muriel made to her husband late that night when he came off duty was, 'Have there been any reports of an aircraft coming down in the estuary?' Sam had heard nothing of any such incident. Again on the Monday evening when he returned home, she asked the same question. No, he had heard nothing. However, the answer came on the Tuesday morning, when the newspaper headlines read: 'AMY JOHNSON MISSING IN THAMES ESTUARY'.

Chapter Two

STRAW HATS AND PIERROTS

s soon as the sixteen-year-old Danish boy, Anders Jorgensen, set foot on the Humber dockside in 1869, he sensed that this was where his future lay. He was an adventurous youth from the small village of Saltofte, near Assens, a lad with the sea in his blood and a strong urge to travel. The Danes were no strangers to Humberside for as far back as the eleventh century their marauding armies had invaded this part of the coastline. Anders arrived to become an apprentice to a Hull smack owner, in spite of the fact that the main source of supply for those wishing to learn the trade was the Anlaby Road workhouse. He had yet to master the English language, something which seemed to come naturally to those of Scandinavian stock, but he had already set his sights on one day becoming the owner and skipper of his own fishing vessel. There was very little steam in those days and the smacks and trawlers, which fished mainly for sole, turbot, cod and haddock, were much resented by the cargo-ship owners, who saw these small ships as a nuisance in the newly opened Albert Dock.

In 1299 King Edward I conferred a Royal Charter upon the town of Hull, giving it the title of King's *Town* upon Hull, the first two parts of its appellation being abbreviated to Kingston. It was eventually to become the centre of the North Sea deep-sea fishing industry and a major seaport, especially after the arrival of the railway in 1840, with its attendant links to the West Riding of Yorkshire and ultimately to Manchester and London. Although the majority of the Hull catches were known for their quantity rather than their quality, it did mean that low-priced fresh fish could now be supplied by rail direct to the south of England, where there was a mass market for this high-protein food.

The rapid increase in the export of fish to the big cities between 1845 and 1872 also coincided with the discovery of the Silver Pits, a deep depression on the sea-bed south of the Dogger Bank, where sole were to be found in great abundance. This caused the number of smacks to increase rapidly in this period from a few dozen, to over 300, mainly as the result of an invasion of the red-sailed, single-masted vessels from Brixham. The Devonian fishermen, some of whom had already moved along the south coast to settle in Ramsgate, now realised that the fish stocks in the English Channel and the southern North Sea were depleting, and sailed north for Hull and Grimsby where the living was more rewarding. Their appearance in Hull and the burr of the Devon accent on Humberside was not particularly welcomed by the local fishermen, and scuffles amongst the men on the dockside were not uncommon. However, much as they were resented, the colony of southern 'foreigners' was to remain.

By 1872 Anders Jorgensen had taken out British citizenship to become Andrew Johnson, and it was not long before he met and fell in love with Mary Ann Holmes, the pretty stepdaughter of one of the Brixham smack owners. There was a short courtship and they were married at Holy Trinity Church on 5 January 1874. It was a real love match, and in later years Andrew Johnson's family and friends would often hear him attribute his success in business to the fact that he had the good fortune to choose an English girl for a wife. The couple had fourteen children during their fifty-six years of marriage, ten of whom survived. There were six sons, all of whom entered the fishery business or had some connection with the sea, whilst their four daughters eventually married and settled nearby. Within a few years the stocky, goatee-bearded Dane had become the successful owner of two vessels, the *Flower of the Harvest* and the *Flower of the Valley*; moreover, he had also patented an important modification to the trawler otter board, more commonly known today as trawl doors, a device which kept the mouth of the trawl-net open.

None knew better than Mary Johnson of the dangers encountered by a skipper or deckhand sailing in the dangerous waters of the North Sea. She knew that in a storm a 30-foot high wave could, and often did, sink a small ship and its twenty crew members in an instant and without trace. Those treacherous seas were known and feared by the fishermen's wives for good reason as 'the widow-maker'. Mary's husband began to recognise the dangers of life at sea more and more, now that he had the responsibility of a large family. This nagging insecurity was no doubt just one of the factors which caused him to look for a shore-based career in the docks. Another was that the days of sail were nearing their end as the more expensive steam-

The initials of the names Andrew Johnson and Knudtzon survive in Neptune Street, near the Albert Dock in Hull, although now part of Andrew Marr International Ltd. (Peter Little)

powered trawlers began to take over. This meant that the single owner could no longer compete against the larger trawler firms which were moving into the docks, forcing the smack owners to join together in business partnerships, if they were to survive. As a result, shrewd, twenty-nine-year-old Andrew Johnson linked up with a Norwegian by the name of Knudtzon, to establish the highly successful import and export company of AJK & Co. Ltd in St Andrew's Dock. Fresh salmon and herring were imported from the Scandinavian countries, the bulk of the latter to be processed into kippers, whilst local fish were salted and cured for export and home consumption.

Shy and petite twenty-two-year-old Amy Hodge, known to family and friends alike as Ciss, lived in what one would term 'genteel poverty' in one of the more fashionable and imposing Victorian houses in Coltman Street, just north of the Hessle Road. The family's near neighbours included solicitors, bankers, merchantmen, engineers and those described simply as 'gentlemen'. There were a few smack owners resident in the road who, with their 'new money' were no doubt looked upon rather scornfully as *arriviste*.

The Hodge family had known better times, but had slid down the social scale somewhat over the years. Ciss's grandfather, William Hodge, was a wealthy mill-owner who had held high office in the City of Hull, being appointed Sherriff in 1859 and Mayor in the following year. He was an ardent chapel goer of the Methodist persuasion, one who took John Wesley's dictum, 'Earn as much as you can, save as much you can, give as much as you can', very much to heart – perhaps too much so, as he was over-generous in his support of much of the chapel-building that took place in Hull during the latter half of the nineteenth century, when a total of twenty-four were opened.

Unfortunately, Will Hodge's philanthropic nature and lack of business acumen finally led to the family's move from Newington Hall,* a seven bedroomed mock-Graeco mansion which stood in four acres of its own grounds on the outskirts of the city, to the less prosperous area of Coltman Street. After his death matters did not improve, for his eldest son, also a William, was an irresponsible charmer who gradually sold off his father's corn mills and spent the money on extensive world travel, leaving his wife to rear the children alone. There can be little doubt that this unhappy family situation left a legacy of insecurity amongst the children, particularly with his musically talented daughter, Ciss.

Will Johnson, the eldest son of Andrew and Mary, married Ciss Hodge in the late summer of 1902. They then rented a small, single-bayed terraced house at 154 St George's Road. The red-brick Victorian property with its minuscule front garden lay south of the railway line between the Anlaby Road and Hessle Road, in an area which came to be known as 'A village within a city' – a close-knit community of fisherfolk with its own culture and customs.

On 1 July 1903 Amy, the first of their four daughters, was born, and almost eighteen months later, a second baby girl arrived, named Irene, or Reenie as she became known in the family. The two girls were to become inseparable

* Newington Hall was demolished in 1908. Coincidentally, the present day Amy Johnson School, situated just south of the Anlaby Road, is quite near to the site of William Hodge's former residence.

Amy Johnson was born at 154 St George's Road, Hull, on 1 July 1903.
(Chris Ketchell)

playmates and Amy recalls in her memoirs, *Myself When Young*, that one of her earliest memories was of the two of them enjoying one of their favourite indoor games, that of leaping about mischievously in their front room from chair to sofa, to the top of the piano, without touching the floor. It was a prank that did little to enhance the appearance of the cushions and upholstery, and one which on doubt earned the strong disapproval of their house-proud mother.

Amy remembers that by the time she was five, she came to realise how pretty her mother was, 'in her bright red and white checked blouse, sleeves rolled up to the elbow, slender fingers and small brown arms buried in foamy suds' as she scrubbed their clothes at the open window of the tiny scullery. One could always be sure that the aroma of fresh-baked bread would be wafting from the Johnsons' kitchen, for to buy shop bread or cakes was much frowned upon. When her housework was completed Amy's mother would gather the sisters around the piano, where she would play and sing to them. Quite often she would play humorous songs, such as 'John Brown's Baby Has a Cold Upon His Chest', and the whole family would join in singing lustily. On other occasions, she would play hymns that they had learned in the St George's Road Wesleyan chapel nearby. It was where the whole family worshipped and where Ciss usually played the organ on a Sunday.

Amy recalls how her father would romp with her on the drawing-room floor pretending to be a bear; and how he would encourage her to play with trains instead of her dolls – not an uncommon practice for fathers who only had daughters. When it came to keeping the two sisters quiet whilst they sat through the service in chapel, their father had a method of pacifying them which met with their full approval. Intermittently, during the service, he would slide his hand into his jacket pocket and pass them chocolate sweets wrapped in pretty silver and mauve papers. He always seemed to have an unlimited supply, so much so, that the longer the local Methodist preacher's sermon lasted, the better the two girls liked it!

Amy, Mollie and Irene Johnson, c. 1914. (The Amy Johnson Estate)

Will Johnson was very much in the mould of his father, being stocky and short in stature, and possessing the same adventurous spirit as the Dane. A few years before he had met Ciss, he had left his regular job with AJK to sail from Hull with three of his young friends to north-west Canada.

At the turn of the century, emigration was attracting hundreds of thousands of young English, Scottish and Welsh people to the colonies of

Australia and Canada. The Irish had already emigrated in large numbers, mainly to the United States, following the collapse of their agriculture which culminated in the Potato Famine of 1845, and between 1850 and 1900 the population of Ireland had fallen from eight to five million. Canada, with its well-publicised cry of 'Free land on the prairie', was now the great attraction for many young Britons, including two of the younger Johnson brothers, Tom and Stanley, both of whom eventually settled permanently in British Columbia. Amy's father and his friends sought a quick fortune by travelling to the Yukon, where the Klondyke gold rush was then at its fever pitch, but after two years, the four men returned home with nothing more than a few small gold nuggets to show for their labours.

The time came for Amy's education, and from the age of five she was sent to a succession of private schools, none of which were of a high academic standard. The headmistresses of these establishments were usually spinsters of good family, women who could scrape enough capital together to purchase a property large enough to accommodate fifty or sixty pupils. They would then employ impecunious elderly spinsters of doubtful ability as teachers. Gullible parents were inveigled into trusting these questionable schools with their young daughters, without realising that the superficial learning acquired in the arts and sciences did little more than ensure that they were distanced from the socially inferior lower classes, and that they were sufficiently educated to attract a husband. At that time there were very few professions open to women of the middle classes, and the main concern of a parent was that their daughter married well. For a young woman to remain a spinster was to admit to being a social failure, as well as being a reproach to her family.

Amy Johnson had a natural thirst for knowledge, despite the poor teaching that she was receiving in these private schools. She was of above average intelligence and an avid reader, a habit which enabled her to surmount and survive the boredom she was forced to endure. She admits to receiving 'a queer mixture of elementary and advanced subjects in physiology, algebra, geometry, trigonometry and biology', but in most subjects it only amounted to 'a smattering'. There seems to have been little strict adherence to age structure within classes in these schools, and Amy remembers lording it over girls much older than herself, as well as playing the role of elder sister to Irene.

Amy was twelve before she and her inseparable sister finished at the

Amy at the time she attended Eversleigh House School on the Anlaby Road, in Hull, c. 1912. (The Amy Johnson Estate)

Eversleigh House School at the end of the summer term of 1915. It was a preparatory establishment on the Anlaby Road, whose Principal, Miss Ada Knowles, appeared to have an arrangement to direct most of her pupils into the Boulevard Municipal Secondary School. During this period of Amy and Irene's education the family moved, as many of those who were striving for improved social standing did, to one of the more elegant houses in the tree-lined Boulevard – a move most probably made in anticipation of their daughters' attendance at the secondary school. They only remained there for a brief period, before moving to an even roomier house in Alliance Avenue. It was here that the Johnsons' third daughter, Mollie, was born two years later.

The First World War brought many changes to Hull, as local men were called to the colours to be shipped off to Europe, where many of them were slaughtered in the muddy trenches of Flanders and other equally dangerous areas of conflict. Most of the men were recruited into the four battalions of the East Yorkshire Regiment, where they were hastily trained at the nearby garrison on Hedon airfield, before being sent overseas. Whilst the regiment escaped the worst of the bloodletting at Gallipoli, the Somme and Passchendaele, they did suffer heavy losses at Serre in 1916, (17 officers and 410 other ranks killed) and again at Oppy Wood, near Arras, in 1917, (34 officers and 770 other ranks).

Two of Amy's favourite uncles were amongst the 7000 Hull men who gave their lives in the war. Uncle Bert Hodge was killed in the big tank battle at Cambrai in 1917, and Uncle Hamlyn Petrie, husband of Will's sister Evelyn, perished a year later. Will Johnson escaped conscription when it became compulsory in January 1916, on two counts: first on age, for by then he was almost forty; and secondly, because he was employed in a vital food industry. As the Marketing and Sales Manager for AJK, he was making regular trips to London for the frequently held Ministry of Food meetings, as well as being on various committees, such as that of the Hull Fishing Vessel Owners' Association.

Hull's civilian population also began to suffer when Germany's Zeppelins mounted their attacks on many of Britain's major cities in January 1915. These huge, cigar-shaped airships, with their powerful Maybach engines, came by night from their bases at Hage and Nordholz in northern Germany, and would unload their bombs onto the docks and houses of the Humberside ports. An attack by Zeppelin L9 on the night of 16 June 1915 destroyed forty houses and shops and killed twenty-four people, with the result that mobs went on the rampage the next morning to attack shops owned by anyone with a German-sounding name.

Winston Churchill, the First Lord of the Admiralty, described the Zeppelins as 'these enormous bladders of combustible and explosive gas', and he thought they would 'prove to be easily destructible'. His optimism proved to be misplaced, for apart from some ineffective gunfire from the ground defences, there seemed to be little in the way of a deterrent against these high-flying night intruders. Night flying was then in its infancy, which meant that at this early stage in the war very few fighter pilots had the necessary experience to intercept the behemoths. Another factor was that the incendiary ammunition which was needed to ignite and destroy them had not then been developed. As a consequence none of the Zeppelins over Humberside was ever recorded as having been shot down. The schoolchildren in Germany jubilantly sang a little ditty, allegedly composed by a relative of the Kaiser, which went as follows: *'Zeppelin flieg! Hilf uns im Krieg. Fliege nach England! England wird abgebrannt! Zeppelin, flieg!'* (Fly, Zeppelin! Help us in the war. Fly to England! England shall be destroyed with fire!)

For Amy these enemy night raids were more a source of excitement than anything to be feared. When the air-raid warning sounded, (the term 'buzzer' was used by Humberside folk), the family would take shelter and the girls would stay up into the early hours drinking hot chocolate and playing games when they ought to have been in bed. Amy recalls in her memoirs an incident during this time:

> One of the highlights of the air-raids is of my father chasing me when I escaped outside, because I wanted to see the Zeppelins and, as I ran hastily indoors again, my sister – frightened at the noise – collapsed into the coal bucket in the darkness of the cellar in which the family were sheltering. I shall never forget the scare this clattering gave to us all, with nerves strained as they were to breaking point after a raid every night for a week. That same night several bombs dropped in our street, but we all escaped harm.

Apart from the inconvenience of the Zeppelin raids during the early part of the war, and the effects of food rationing, life went on in Hull very much as before. There were some minor changes. Women were filling the jobs left vacant by the men, which meant that the plum-and-white-liveried trams which swayed and clanged along the Hessle Road to Pickering Park – a favourite halfpenny ride for Amy and Irene – were now manned by lady conductresses. And as women went into the factories on essential war work, Ciss Johnson fussed and fretted, as she found it more and more difficult to employ 'the little maids' that had been so plentiful before the war. Ciss was a born worrier, who constantly suffered from 'nerves'. She tended to be over-concerned in the parenting of her daughters and it might well be that she passed her own childhood insecurities on to them.

Apart from the irritation of the black-out regulations, which the police strictly imposed, entertainment in Hull was never seriously curtailed during the hostilities. The theatres and cinemas remained open very much as before, and from a study of Will Johnson's meticulously kept diaries the family made ample use of them. Amy became a life-long cineaste right from an early age when, as a treat, her father would take her with him on a Friday evening to one of the local cinemas. No doubt when she was old enough to go unaccompanied, he steered her away from the more dubious picture palaces, such as the Magnet on the West Dock Avenue with its rough wooden seats, or the Eureka in the Dairycoates area, which had a reputation of being a 'flea-pit' and was known by the locals as the 'Laugh and Scratch'.

Amy loved to watch her screen idols, favourites such as 'America's Sweetheart', Mary Pickford, as she escaped the clutches of the villain or the titian-haired goddess, Clara Bow, seductively embracing her man of the moment. Seated in the fantasy world of the darkened auditorium, Amy would roar with laughter at the antics of the baggy-trousered, cane-swinging Charlie Chaplin or sit with bated breath as the hapless Harold Lloyd hung by his fingertips from the window ledge of some New York skyscraper. Little did she realise that not only would she one day meet these Hollywood stars, but she too would be recognised by the public as a household name. The silent movies with their subtitles must have made a great impression on Amy for her letters in later life are heavily sprinkled with words like 'cos' and 'tho', abbreviations prevalent in conveying the silent-screen stories. It was not uncommon in those days for poorly educated parents to be heard whispering to their more literate offspring, asking for the meaning of the words appearing on the screen.

On the morning of 24 August 1915 Ciss Johnson accompanied her two daughters for their enrolment at the Boulevard School. Amy and Irene sat nervously alongside other pupils in the large, sparsely furnished hall to await their turn for the interview with the Headmistress, each of them slightly embarrassed as they wore their school blazers for the very first time. The Boulevard uniform for girls consisted of a maroon blazer with a silver-grey edging, worn over a matching tunic with white blouse and maroon tie. The straw boater, or 'banger' as it was known to the older girls, was the summer headgear, whilst in the winter a tight-fitting cloche-type hat defied the strong east winds which blew off the North Sea.

Ciss was more than a little disturbed when she was told that her girls' entrance papers had been mislaid, and was asked to come back at a later date. According to Amy the problem was finally resolved when she was placed in a class with girls a year younger than herself, as there was no room for her in a class of her own age. It seems far more likely that her academic ability was judged to be more appropriate to those of the first-year intake, because she had spent a year longer at the Eversleigh than she should have done. It might well have been that the Headmistress was prepared to review her progress after a while and place her in her own age group, if it was warranted. However, as it transpired, Amy was always a year behind at the Boulevard, which meant that she was to stay on at the school until she was nineteen.

Being a year older than the rest of the class, Amy soon found that she could more than hold her own academically, a situation which unfortunately made her rather lazy. However, it did enable her to assume a leadership role

The Boulevard Municipal Secondary School (now 'Rosedale Mansions') where Amy and Irene enrolled in 1915. (Chris Ketchell)

which she never quite lost throughout her school career. She excelled as an athlete, with displays of daring on the trapeze and high spring-board jumping in the gymnasium. She was also a fine swimmer and hockey player. Boulevard was a coeducational school, although the classes were segregated until the pupils reached the upper school. Not having any brothers, Amy admitted to taking a great interest in the opposite sex, and was on one occasion reprimanded for flouting school rules by being seen out walking with a boy in the local park when they were both in school uniform – a practice which was strictly forbidden.

Unlike Irene, Amy was a natural tomboy, a trait which was soon noticed and admired by boys of her own age. She would join in their cricket matches and quickly gained a reputation as the only girl in the school who could bowl overarm, *and* with some accuracy. Her involvement in what was considered to be a masculine sport did, however, result in a rather nasty accident, when she took a fast cricket ball full in the face, breaking one of her front teeth. This was a great setback for Amy, for in those days dentistry had not developed the skills that are available today. The impairment also left her with a slight lisp which, much to her annoyance, was a matter for much merriment amongst the boys. As a result, for a number of years whilst she was at the Boulevard, this disfigurement caused her to become withdrawn and for some time she was rarely seen to smile. Repeated visits to the local dentist and the fitting of a plate eventually restored her confidence, but it was a problem which was to bother her well into later life.

The adventurous and rebellious side of Amy's nature became more and more evident as she neared the sixth form, and it was around this period that she tried to organise a swimming event for girls at the Boulevard. Unfortunately, the rather austere and pompous Headmaster, Frederick De Velling, a bespectacled academic whose short-sightedness was not only of the ocular kind, had very fixed ideas on what constituted a suitable sport for young ladies, and swimming was not one of them!

Amy was not deterred by her headmaster's disapproval and she set about manoeuvring one of her more pliable teachers into backing her plan for holding a swimming contest for the girls at the local baths. Both Amy and Irene had already cajoled their father into paying for the hire of the swimming-pool, whilst Amy's French teacher, Miss Sheppard, had agreed to present the prizes on condition that the event was not given any publicity.

Mary Sheppard had joined the school as a twenty-four-year-old at the outbreak of the First World War and soon came to a similar opinion of the Headmaster to Amy, for she recalls that 'Mr De Velling taught me a sound method of pleasing him. I learned to agree with his ideas completely or keep silent.' On the occasion of Amy's unofficial swimming event she kept silent. The Boulevard's historian, John Hicks, recalling Miss Sheppard's long career at the school, states: 'No teacher at the school has ever been so popular, or had so many old pupils as visitors during a long retirement. She died in January 1990 just a few weeks short of her hundredth birthday.'

The illegal swimming contest was not the only example of the stubbornness and determination of character evident during Amy's time at the Boulevard School. When Irene left the school in the summer of 1920, her parents decided that she should finish her education at the Hull High School for Girls, in Spring Street. Not only was this school of a higher educational standard than the Boulevard but, more importantly as far as the fashion-conscious Amy was concerned, their pupils wore a more stylish headgear. Amy was quite envious of Irene's new panama hat, so much so that she became determined to introduce them at the Boulevard. After discussing her proposal with other girls who also detested the 'bangers' that sat so uncomfortably on their heads, Amy suggested that a number of them purchase panamas, stitch on the old Boulevard headband and badge, and appear in them at the school on a certain day.

When the day for the 'Rebellion of the Straw Hat Brigade' finally arrived,

Beverley Road Swimming Baths, scene of the unofficial swimming gala, where Amy persuaded Miss Sheppard to adjudicate. (Chris Ketchell)

Amy found, much to her consternation that she was the only pupil wearing a panama hat! Inevitably, as soon as she entered the school she was spotted, not only by her peers but more importantly by the eagle eye of the Headmistress, Miss Edith Clews. She was a small woman who, in an attempt to offset her lack of inches, wore her hair in an upswept bun, and she brooked no nonsense where discipline was concerned. She had an air of self-importance which was most probably nurtured by the fact that although she held the title of Headmistress, she had lost status a few years earlier when she was effectively downgraded as merely a senior mistress under De Velling.

Miss Clews' method of punishment for such an action as Amy's was to inflict humiliation. This meant that the rebellious pupil was made to endure an interminable wait of several hours outside her door, whilst fellow pupils passed by, suppressing their giggles in cupped hands. Amy recalls that 'almost everyone in the school had the chance to pass me and twit me for being in such an undignified position'. There is no doubt that her action, foolish as it was, required a great deal of bravado and a steely resolve to carry it out. These were characteristics that were eventually to bring her worldwide acclaim.

One wonders just when and where Amy's interest in aviation first began, for whilst at school she had only seen aeroplanes portrayed in news items at the cinema. After the cricket-ball incident and the jeering that she had taken from the boys, she withdrew into a world of her own, shunning team games and becoming very much a loner. This period coincided with her acquisition of a bicycle, a present from her parents, which meant that for the first time she could explore and enjoy the freedom of the surrounding countryside. She admitted that these long, solitary rides were not confined only to weekends, or after school times, but that she would often play truant from school. On one occasion she cycled out to a nearby airfield, most probably the one at Hedon, but failed to see any flying activity. The Blackburn Aircraft Company, on the north bank of the River Humber at Brough, which produced seaplanes and flying-boats during the First World War, was also only a few miles west of Hull, but being a military establishment, it would have had restricted access to inquisitive young ladies on bicycles.

At other times Amy and the whole family would cycle north of the city centre up the Beverley Road to visit her aunt and uncle at the nearby village of Cottingham. Will's sister Annie had married John Campey, a keen chess player who would invariably challenge his brother-in-law to a game almost as soon as he had set foot in the door. Amy and her sisters lived in a warm, close-knit family atmosphere with a constant stream of aunts, uncles, grandparents and friends visiting their home in Alliance Avenue.

In the days before radio or television, home entertainment was restricted to sing-songs around the piano or listening to the hiss and scratch of records on the wind-up gramophone. Alternatively, cards, dominoes, ludo, chess, draughts and other board games were firm favourites.

During the summer the Johnsons rented a house near to the beach in Trinity Grove, Bridlington, for as long as three months at a stretch. This meant that for some of that period, Will, Amy and Irene commuted by train between Hull and their beloved 'Brid'. (In later life Amy was known to tell people that she originated from Bridlington, rather than admit that she came from Hull.) The girls would spend glorious hours on the beach, with Mollie paddling at the water's edge or being chaperoned on her donkey rides by a none-too-willing older sister, whilst the more daring eldest daughter delighted in scaring the wits out of her mother by swimming further out from the beach than was safe, causing Ciss to remonstrate, 'Don't go out so far Amy, you know it's dangerous.' The evenings were spent either watching the pierrots on the Spa, or going to the pictures at the Pavilion or the Palace. On Sundays, Will Johnson would take the girls to worship at the Methodist chapel, whilst Ciss would invariably be asked to play the organ at the local Baptist church.

The war appears to have been a prosperous time for the AJK firm, and for Will Johnson in particular, for towards the end of 1918 Amy's family moved into a large, semi-detached Edwardian house in one of the better middle-class residential districts of Hull, 85 Park Avenue, and it was here on 30 May 1919 that their fourth daughter, Betty Margaret, was born. Perhaps it was a final attempt to produce a son and heir after three successive daughters. If so it was a disappointment for Will Johnson, but the new baby girl was loved and fussed over by the whole family just the same, and came to be known affectionately as 'Bee'.

As the Johnsons became more affluent they purchased one of Lord Nuffield's popular bull-nose Morris Oxfords, and then went on to become a two-car family when they added a Standard saloon, so that Ciss could have her own vehicle. However, at the immediate end of the war the only means of transport in the household was a motorcycle, around which there is an apocryphal story relating to Amy. In his book, *Queen of the Air*, Gordon Snell records that one day Amy rode her father's machine without his permission. She is said to have set off down Park Avenue making U-turns at each end of the road and passing the house shouting, 'How do I stop it?' Her worried father came to the front gate gesticulating instructions on how she should do so, but to no avail. Eventually, she either stalled the machine to a halt or ran out of petrol. The author's source for this incident was most probably Amy's sister Mollie, since she is the only family member acknowledged in the book as a contributor.

Amy tells us in her memoirs that she was forced to stay on at school until she was nineteen and spent two years in the top form in order to allow her hair to grow! This is an unlikely story, for although she admits to causing her mother much grief by cutting off her two long plaits in an act of rebellion, the true reason is more likely to have been the mix-up at the Eversleigh School and her initial enrolment at the Boulevard. The new fashion of 'bobbing' the hair, so much favoured by young women in the twenties, had not gone unnoticed by Amy, but it was still the conviction of her parents that young ladies should be able to put their hair up.

As Amy reached the end of her schooling in 1922, having gained the necessary qualifications that would take her to university, she was confronted with the need to make a career choice. She had no clear idea of what she wanted to do with her life, although she looked forward to enjoying the freedom of being at university. One of the main influences in her choice of teaching as a vocation was the fact that the Headmaster's daughter, Lizzie De Velling, who had been a star pupil at the Boulevard and had passed the Cambridge Senior Examination before going to Sheffield

University, had returned to teach at her old school. Her achievement automatically became an example for other pupils to follow and as a result Amy was strongly urged by the staff and her parents to follow suit. She was easily persuaded and went along with the advice, although the idea of teaching did not really excite her.

Chapter Three

MRS PETRIE'S INTRODUCTION

Amy's total experience of romance had so far been limited to occasional flirting with some of the boys at the Boulevard School, and to what she had seen on the silver screen. It was not until the latter part of 1921 that the eighteen-year-old was to fall hopelessly in love.

It was the custom for Amy to accompany her parents and sisters on their usual extended summer holiday to Bridlington, and as in previous years, there was the usual family get-together of aunts, uncles and grandparents. On this particular occasion one of Amy's favourite aunts, Evelyn Petrie, was present. She was one of Will Johnson's younger sisters who, since being left a war widow when her husband was killed in France, had gone with her young son to live at her parents' home in Bridlington. Evelyn was a keen tennis player and a member of the local club and during that summer she encouraged her niece to play. Amy showed little interest in the sport at first, until she was invited by her aunt to attend a party being given by the club.

One of the guests at the tennis club's gathering was a twenty-seven-year-old Swiss businessman, who had arrived in England earlier in the year to work for a Swiss company based in Hull. Hans Arregger could be described as no more than faintly handsome, having a rugged kind of masculinity. He was of medium build with short, thick hair and clean-cut features under a strong brow. One immediately noticed his quizzical grey eyes, which could be penetratingly severe when serious, or twinkle with amusement when the mood suited. Moreover, he had the cool, self-assured manner of a man of the world, typified by the way he smoked his oval Turkish cigarettes from an elegant holder. His command of spoken English was quite good, although

These tennis courts at Beaconsfield Public Gardens, Bridlington, (circa 1935), would have been familiar to Amy once she took up the sport since the Waterloo Pierrots performed quite near to them on the beach after WW1.
(East Riding Library and Information Services)

there was the occasionally misplaced word, something which came across to the listener as rather appealing.

None of this was missed by Amy, who was smitten as soon as her Aunt Evelyn introduced her to him. Initially, he took very little notice of her, for to him she was no more than a schoolgirl, but she was already chalking her cue, puzzling over how she could heighten his interest. She used several methods. Her immediate priority now was to become an enthusiastic tennis player, and then to find out what his interests were and to share them.

It was most convenient for Amy since she was already studying French in the sixth form, and it was also one of the subjects which would form part of her studies at university, for Hans was anxious to improve his own skill in that language. As a consequence, it was not long before Amy was accompanying him on his regular visits to the French Circle and to other clubs in Hull in which he had an interest, such as the Literary and Philosophical Society. They were both very much into reading and she started to share his literary interests when he introduced her to classical writers such as Arnold Bennett, George Bernard Shaw, Henrik Ibsen, Johan Strindberg and other authors she had not previously read.

Whilst their relationship was initially just on a friendship basis as far as Hans was concerned, there is no doubt that Amy was intending it to be a more serious one. And as he spent more time with her, he found himself being attracted to her as the woman rather than the schoolgirl whom he had first met. Her dimpled cheeks and the seductive blue-grey eyes were hard to resist, and before long Hans was taking her to dances or to the local cinemas. At weekends they would go for long rides on his motorcycle into the surrounding countryside, with Amy holding on tightly behind him on the 'flapper bracket', enjoying the thrill of speeding along country lanes with the wind in her hair. That summer of 1922 was all too short for Mrs Petrie's niece, and as the time drew near for her to leave for university, she wondered just how she could maintain and develop her new friend's interest whilst she was away.

During the six-and-a-half-year period of their relationship, Amy and Hans were to conduct an affair in which their most intimate thoughts were exchanged largely through correspondence. Almost three hundred of Amy's

The Literary and Philosophical Society meetings which Amy and Hans Arregger attended were in the Royal Institution in Albion Street, Hull. The Victorian building was destroyed by enemy action in June 1943 and, sadly, its site is now merely a car park. (Hull City Council, Local Studies Library)

letters to Hans have survived and are now in the Hull Central Library, having been sold at auction by the Arregger family in 1982. Inevitably, they are very revealing of her character, and we soon discover that she was a young woman who suffered from frequent and quite drastic mood swings.

Amy's first letter to Hans was written just before she left the Boulevard for university, and in it she sent her apologies for being 'perfectly horrid last night' in front of two of his Swiss friends. She excused herself by blaming the strain of her school exams and attempted to soften the blow by adding the flattering comment that 'foreigners always seem to have such perfect manners and I'm afraid that you'll think I haven't got any'. Although it was a letter of contrition, it did not bode well for her plans of romance.

Amy arrived at Sheffield University early in October 1922 to sit her entrance examinations, a sifting process to verify whether she was up to the standard in French, English, Latin and Modern European History, the subjects which she had chosen. These exams would also determine whether she would be allowed to take an honours or an ordinary degree. Much to her dismay she failed miserably in all subjects and began to envisage the ignominy of returning to Hull in some disgrace. It would also mean that her father would be required to return the teaching grant that she had already received and partially spent. However, rescue came in the form of an interview with the sympathetic dean, who gently steered her towards taking a less demanding course of study in Economics. Amy was quite happy with this second-best arrangement since she had no real desire to teach, and as a last resort had toyed with the idea of a job in the Civil Service. However, her real aim was to travel to France, and now that she had met Hans, maybe to Switzerland as well.

The halls of residence at the university were oversubscribed, which meant that 'freshers' such as Amy and her two colleagues, Gwyneth Roulston and Miriam Benham, also from the Boulevard on teachers' grants, had to find private lodgings in the Sharrow district of the city. Almost immediately Amy was writing to Hans telling him that she was having the time of her life. 'We've got the gramophone on tonight and all I want is someone to dance with. I'm simply aching for a dance.' She would then tease him by telling him how she had filled her dance programme with requests, not from freshers like herself, but from the senior men!

Amy's time in Sheffield coincided with a period of great social upheaval in post-war Britain as people - young women in particular - began to throw off the restraints of the Victorian and Edwardian eras. Of necessity, women had undertaken men's duties during the war and had acquitted themselves as their equals in many of the jobs. In the factories they had worked as machine operators and mechanics, whilst in transport they had begun driving trams, buses and ambulances. Furthermore, the suffragettes had won the vote for women in 1918, albeit only for those over thirty (full and equal rights did not materialise until 1928), and the old pre-war barriers had now gone forever.

Single and young married women now presented a more brazen image, using lipstick, mascara and nail-varnish. Eyebrows were plucked, faces whitened and rouge applied to highlight the cheekbones; make-up, once thought to be the sole province of the *demi-monde*, was now commonplace. Women dyed and bobbed their hair and smoked cigarettes openly in public, whilst chaperons became a thing of the past as more and more women lived independently at university, or in the big cities as they took secretarial and office jobs. They enjoyed a freedom quite unknown in their parents' day. Even dress fashions reflected this emancipation by becoming more daring and overtly sexual. The felt cloche hat and the low-waisted dress cut to above-knee length became popular, as busts were flattened by the liberty-bodice to give a straight-line boyish figure. Those girls more generously endowed by nature were forced to pull their bodice strings just that little bit tighter!

This period was also accompanied by the dancing craze which swept the Western world after the war, every bit as virulent as the Spanish 'flu had been before it. Not only did it attract the young bead and tassle-swinging jazz babies, but men and women of all ages. The old-fashioned waltz now gave way to new and more exciting dances coming from across the Atlantic, such as the fox-trot, the Black Bottom, the Charleston and the exotic tango. The smoke-filled nightclubs and dance halls thrived as the 'bright young things' and their war-weary partners took to the floor. Even in the most modest of homes the lounge carpet would be rolled back and the gramophone wound up, as couples slid their feet across the linoleum to the syncopated strains of music from dance bands such as the Original Dixieland Jazz Band and their like.

Amy, Gwyneth and Miriam were now part of that scene in their own modest way. There is no doubt that this dancing craze on both sides of the Atlantic, described by some as an epidemic, was a reaction to the sickening horrors of the muddy trench warfare which had claimed the lives of almost three quarters of a million young British men.

Before Amy left home she promised her parents that she would continue with her chapel-going, and in one of her early letters to Hans she reminded him that she was keeping to her word. However, her sole experience of a social evening given by the university's Christian Union was not a happy one. After 'patiently sitting through two supposedly humorous songs' she discreetly made for the door, only to be met on the stairs by the young secretary of the Union who rather forcibly persuaded her to return.

The entertainments continued until a break in the proceedings, when 'musical chairs' was announced. Amy recalled, 'That just put the tin hat on it – I mean, that added the final straw to an already unsupportable situation. Seizing a favourable opportunity when my guardian's eye was in another direction I ran for the staircase. Hastily descending it I threw myself into my hat and coat, and heaving a fervent sigh of relief I left the building.' She immediately made a beeline for the local cinema, vowing never to return to the CU. Unlike her mother, with her strong Nonconformist faith, religion seems to have made only a superficial impact upon the young fresher from Hull.

Amy's first ploy in encouraging a none-too-willing boyfriend to write to her regularly, was to help him improve his written English by promising to correct any mistakes he might make. She in turn would write back in French and he similarly would correct her mistakes. At the same time she attempted to establish a non-threatening relationship with him by signing herself '*votre soeur adoptée*', and by telling him that she desired nothing more than 'a brother's affection'.

The 'tutorial sessions' which were conducted between them were short-lived, for he would return her letters with corrections marked copiously in red ink, commenting, 'This is all wrong,' or, 'This is rubbish.' Such comments caused Amy's hackles to rise and she would come down heavily on his clumsy mistakes, such as, 'You might correct me this letter', scolding him that it should read, 'You might correct this letter for me.' Not surprisingly, it was not long before Amy was telling Hans that it was too much bother for her to continue to write to him in French, and that she would try again when she had become more expert in the language.

As time progressed, Amy became bolder and began to use less subtle tactics when she told Hans that her friend Miriam thought that she needed steadying down by becoming engaged. On another occasion she pushed the idea of an engagement even further when he mentioned that his landlady had told him that Amy was welcome to come and have tea at his lodgings when she returned at Christmas. Amy saw this as an opportunity to write, 'I hope you've enlightened Mrs Baker as to "our engagement". The only engagement we've got is to write to each other, isn't it, *cher frère*?' Exactly how Hans reacted to these suggestions, other than with mild amusement, we have no way of telling. One gathers that he was not a man to be hurried into any decision; in fact he appears to be have been a man of extreme caution.

In her memoirs, *Myself When Young*, Amy admitted that at university she 'did all the usual things, living a life of dances, games, picinics and parties, interspersed with occasional lectures. Examinations were passed by the barest margin, and then only by dint of cramming for the week beforehand.' However, when she returned to Sheffield after the Christmas recess, she found that last-minute cramming was not enough, for she told Hans that she had failed miserably in French and that her future now looked bleak. She continued by thanking him for the box of chocolates he had sent her and at the same time tried to make him feel guilty about the inertia of their relationship by saying, 'It was no trouble at all in correcting your letters, but I feel disappointed to think that you only wrote to me to have your English improved.'

It is obvious that Amy's armoury was not limited to only one weapon, and if Cupid's arrows were falling short of their mark, she was only too happy to improve his aim.

Amy was often homesick during her first term at Sheffield and as a consequence she was always delighted whenever Irene offered to visit her, usually for a few days at a time. The two sisters enjoyed a close relationship, although there was always a slight tinge of jealousy on Amy's part. Admittedly, there was never any real spite in the rivalry, but it was there just the same. This came about primarily because Amy felt that Irene was the favoured daughter of the family and that she was spoilt, particularly by her mother. Irene could always get her own way where her parents were concerned. If she was spoilt, then it most probably came about as a result of her suffering a nervous breakdown during the time she was training to be a teacher. On that occasion she had a bout of depression and frightened her parents by threatening to commit suicide. The other possible cause of Amy's jealousy was that Irene was the prettier and most probably the more talented of the two sisters.

Irene had been gifted with a good singing voice and she also had an inescapable charm which made her attractive, particularly to the opposite sex. However, despite their differences the two sisters enjoyed each other's company and spent a part of the summer of 1923 holidaying together on the Isle of Man. Hans had been busily engrossed at that time forming a small business in partnership with a fellow Swiss by the name of Raymonde. The two men rented a warehouse and offices in a run-down area of the Albert Docks, adjacent to the Victoria Pier, where they dealt in the import and distribution of fruit, vegetables and vegetable oil. This new venture was to become a constant bone of contention between Amy and Hans, as it took up more and more of his time, both at home and abroad.

The Christmas of 1923 was a time of immense excitement in the Johnson household; the whole of Amy's family came together in a grand reunion to celebrate her grandparents' golden wedding anniversary. It was a unique gathering of the clan, with two of Amy's uncles and their families making the arduous 8000-mile trip from Canada. Uncle Tom was the managing director of the Canadian Cold Storage Co. in Vancouver, whilst Uncle Stanley, also from British Columbia, had prospered in the fishery business. The celebration took place on Christmas Eve in the banqueting room of the City Hall in Hull, when 150 guests sat down in honour of Andrew and his wife Mary. Amy was one of the fourteen grandchildren present.

The sprightly, seventy-two-year-old founder of the Johnson family, although ostensibly retired, was still active in the firm of AJK, and when interviewed by the local newshound, it became clear that he had lost none of his old sense of humour. The newspaper reported that 'it seemed many years since he could say to one of his children, "Do this", and to another "Do that", and it would be done promptly (sometimes), but when he looked round and saw the size and weight of his ten children, he felt that discretion was the better part of valour. He would like to feel that father's word was still obeyed, and so he would, at the gathering, ask his children to all stand up and then sit down.'

Early in the New Year, Hans moved his lodgings to another part of Hull, and Amy took the opportunity to compare the peccadilloes of their landladies; hers was 'loquacious and tactless' or too inquisitive, whilst in her opinion, his present one was pampering him too much. Amy was to become an expert on the subject of landladies and lodgings, for throughout her life she was destined to live a nomadic existence. She was never gregarious by nature, particularly when it came to sharing rooms with other girls, for then she could be rather choosy. This is borne out by the fact that when she was offered accommodation in one of the university's halls of residence in her second year, she still preferred the freedom of being on her own. And so it was not surprising that she moved four times during her time at university.

During her first term at Sheffield, Amy had dropped hints to Hans to see if he might care to visit her for a weekend, but he was not easily drawn, in spite of her attempts to provoke his jealousy by telling him dubious stories of the wild life that she was leading. Hints in her letters such as, 'It's tennis tomorrow . . . a dance on Thursday', and, 'I am going to stop flirting some time this week', were to no avail until the latter part of 1924, when there is a letter thanking him for having replied to her letter so speedily. Amy saw his eagerness to write back as a sign that he was prepared to take the lead in their relationship.

Amy was required to go to London for an interview during the week before Christmas. The reason for this trip is not clear from her letters, but it did mean that she would need to stay in the capital overnight. She was quick to ask Hans if he would go with her – not an unreasonable request since he often did business in London. Hans seemed to need little persuasion, although Amy made sure he kept to his word with, 'Do you really mean it about coming up to London with me? It sounds far too exciting to be true. Do make the "will" very, very strong, so that you'll find a way.' He had obviously told her that he was extremely busy, but she continues, 'You know, it's a sign of "great industrial activity" to be "hopelessly busy" [no doubt mocking his grandiose use of the language here], but let it rip for just two days.'

In a final reminder to him just before they were due to meet on the platform at Doncaster station, she thanked him with, 'It's awfully good of you to have seen to the booking of the rooms.' She then coyly adds, 'I don't quite know what I'd have done if I'd had to rely on myself.' What remarkable determination she showed in getting her own way! It was a characteristic of Amy Johnson which was to emerge most strongly in the years ahead.

There is no doubt that from this moment on Amy became Hans's mistress, or as he would euphemistically term it, she became his *petite amie*. She was obviously a willing collaborator, although one suspects that Hans had made her more compliant by encouraging her to read the works of writers who advocated the concept of 'free love' and the 'emancipated woman'. It was unusual for middle-class young ladies from the provinces to break the social taboos that were then still in place, and one must conclude that the illicit night that she spent with Hans well illustrates the rebellious and risk-taking side of Amy's nature. Contraception was not then widely available – or reliable for that matter – and pregnancy outside marriage brought with it a stigma and shame almost totally unknown today.

One also needs to take into account the fact that during those early years, Amy was little more than a rather frivolous *ingénue* kicking over the traces. Her experience of life outside the cinema, classroom and lecture hall, was virtually non-existent, whilst Hans, who had served as an officer during the war, was most probably already sexually experienced. When viewed from the morals of that time his behaviour would certainly have been considered caddish, since he had not even given the slightest indication of an engagement, let alone marriage. Maybe he had some excuse, since the 'breach of promise' rule then current, with its financial penalties for reneging bachelors, could be quite severe.

Amy's deeper involvement with Hans meant that her immediate priority was now marriage rather than a teaching career and, as she explains in her memoirs, 1925 was to be her final year at Sheffield. 'By the end of the third year, I was becoming so restless and tired of the life, I begged my father to let me off the fourth year which should have been spent obtaining my teacher's diploma.' Her restlessness is evidenced by a letter to Hans at the beginning of the year, at a time when she was moving yet again to new digs. She wrote, 'Won't you please write me just a few lines . . . I'm feeling most hopelessly depressed.' Try as she would, Hans was no more forthcoming in responding to her less than subtle arm-twisting for an engagement, or to her scheme to accompany him to Switzerland to meet his family.

During her last two years at university, Amy struck up long-lasting friendships with Winifred Irving and Constance Tupholme (who was nicknamed 'Tuppy'). Both were local Sheffield girls studying for degrees, and along with Gwyneth Roulston they became her constant companions, often spending weekends together camping at Baslow in the Derbyshire countryside. Few young women, when in close intimacy with their peers, can resist confiding their secret romances or affairs, and there was great excitement amongst them when Winifred announced that she had recently become engaged to a young Irishman. Amy was not slow in relating the event to Hans and describing Winifred's 'sweet little engagement ring'. She finally added, 'We are all very envious, and have decided we'll all get engaged as well.'

There is no doubt that Winifred Irving, widely travelled by virtue of her involvement with the National Union of Students, helped to broaden Amy's interests considerably. She encouraged her to be far more outgoing and to look beyond the narrow confines of the university dances and weekend hockey matches. One of the ways in which she did this was by inviting Amy to take part in an inter university debate at Bangor in February 1925. Knowing that Amy was not yet confident enough to be one of the main speakers, she asked her to make the summing-up speech. Amy's shyness prevented her from doing so. In view of her subsequent history, there is a strange touch of irony in her comment on the matter to Hans: 'Can you imagine me standing on a platform, facing hundreds of learned students and summing up! No, I'm sure you can't. I can't either.'

Winifred Irving and Amy enjoying themselves on one of their camping weekends at Baslow, Derbyshire. (Patricia Stewart)

However, Winifred did not give up trying and at the Easter vacation she invited Amy to accompany her and her fiancé to the Oxford Congress of the NUS. Winifred was a member of the executive committee and Amy was quite flattered when asked to help out by taking shorthand notes of the meetings and then transcribing them. The only thing about the trip which seemed to worry her was finance. Like her mother, she had the Hodge inability to manage money, a trait which was to dog her throughout her life. Writing from her Oxford lodgings, which she described as 'awfully ripping ones', she told her mother that she got 'dreadfully worried over money matters' and proceeded to plead for a loan of £2.2.0 to pay her rent. She continued by explaining her pecuniary embarrassment in some detail with, 'Every meeting and meal we go to we have to put our things in the cloakroom and pay 2d. Isn't it scandalous? I kept mine on for the meeting.'

Ciss Johnson's worries were not confined solely to those of her eldest offspring. Irene had fallen out with her mother some months earlier and much to the dismay of her parents, was now living in lodgings on the Anlaby Road in Hull. Their highly strung daughter had needed careful handling ever since her nervous breakdown, and more so now that she could not find suitable employment. Amy considered her sister to be something of a scatterbrain with wild romantic notions, especially after she told Amy that she was toying with the idea of going on the stage as a singer and dancer. In one of Amy's letters to Hans she described how Irene had confided to her during one of her visits, that 'a Frenchman in her digs has fallen in love with her and they're going to Paris together soon by aeroplane, and she's going to stay there and live a "wild" Bohemian life. She'll wake up soon.'

Perhaps it is not surprising that in a letter written to her mother around this time Amy wrote, 'Hope the Vibrona will do you good. Isn't Phospherine better? It always does me good.' She then closed with, 'Now be careful with your health.' Both mother and daughter were constantly monitoring each other's state of health, and in a way they were sympathetic to each other's hypochondria.

Amy had enough worries of her own to be too concerned with Irene's problems, for in spite of her need to get down to serious study if she was going to succeed in the final exams, her obsession with Hans after their 'honeymoon' in London was proving a big distraction. Anxious to further their relationship, Amy was quick to invite Hans to the university Carnival Ball, which was scheduled to take place on Shrove Tuesday. She was keen for him to meet her new friends, Winifred and Tuppy, and much to her surprise he readily accepted. She took this as a sign that he was really serious about her and this conviction was reinforced when flowers arrived for her at her digs soon after.

It appears that the weekend did not go as well as Amy would have wished, however, and in a letter written soon after he returned to Hull, she reminded him of the physical intimacy between them by telling him that she was glad that she had not been wearing any flowers on her red frock when they danced, 'because they would probably have got crushed, and I don't like to see flowers spoilt'. Her letter appears to have smoothed things over, since soon after she thanked him for sending her some lovely peaches and primroses.

Amy's pursuit of Hans did not diminish when he decided to spend a holiday in Switzerland visiting his family during May of that year. His widowed mother and three of his four sisters lived in the family home at Lucerne and it just happened that his visit coincided with his birthday. Just

Gwyneth Roulston posing on what was most probably Hans Arregger's motorcycle (a 1924 Triumph, type 'R') during one of his visits to Amy whilst she was at Sheffield University. (Patricia Stewart)

before he left, Amy reminded him of her yearning for him. 'I woke up this morning with the strangest feeling that you were kissing each of my eyelids in turn, as you often do – I don't know whether I'd been dreaming, but it was awfully strange. I've not had such a *certain* feeling like that before.'

Hans was obviously flattered by Amy's dependence upon him and he pursued the relationship by sending her red roses – a token which she was fond of referring to as 'the language of flowers'. There is little doubt that he had a Svengali effect upon her, since she would seek his opinion on almost every aspect of her life. She would ply him with a range of questions such as whether she should go to seek employment in America where her father had contacts, or whether she should apply for a job with Brown & Boverei, a Swiss engineering concern at Baden. What would she do if she failed her degree? However, her immediate concern was to know if he would tell his family of his new English girlfriend, and she hinted that he should do so when she wrote, 'What will your mother think if I write too many letters?'

During Hans's stay in Switzerland Amy moved lodgings yet again, this time to a small cottage in Hathersage, a village some 10 miles outside Sheffield on the edge of the Peak District. Her reason for doing so was that the digs on the busy Glossop Road suffered from too much traffic noise and she could not concentrate on her studies. Whilst she gained from the quiet of the countryside, she quickly discovered other disturbances equally distressing – beetles! In her memoirs, she wrote:

> They are my pet aversion and I have a mad, blind horror of the beastly things. It was sheer ignorance of their obnoxious habits which made it possible for me to continue living there. I laboured under the comforting delusion that they were tied to the ground, that they could not climb or fly. At night-times, lying shivering in bed, I would hear stealthy rustlings behind the walls, but told myself sternly that this noise was caused by mice, of which I am very fond. One evening as I lit the gas in my bedroom, I saw something black run quickly up the wall and escape inside a crack of which there were hundreds in the walls and ceiling. I held my breath, firmly telling myself that this could not be a beetle because it could not climb. Inspired by a sudden idea, I ran downstairs and brought back a roll of adhesive tape which I stuck over all the cracks. I could not bear to go to bed until every crack was covered. I was terrified the 'black thing' would come out.

With less than a month to go before her final exams, and in spite of having failed Latin composition in the Easter exam, Amy's social life was still taking priority over her studies. Hans had accepted Amy's invitation to spend Whitsun weekend at Hathersage and she had arranged for him to stay at one of the two local inns. After reconnoitring both establishments Amy decided he was more suited to the cheaper of the two, not because of the money saved but because the other inn had, according to Winnie, a reputation for being notorious. 'Other people's wives and other people's husbands stay there for weekends. It's not a fit and proper place for you. Winnie advises it, and I'd back her anywhere.' And so it was that Amy guided him to the safer haven and he was spared the more salacious temptations of village life - real or imagined.

During her last two weeks at university Amy struggled with her finals. Her nerves were stretched to the limit, owing to the little time that she had given to her studies, which meant that she was cramming right up until the day of each exam. It was a case of blindly memorising those parts of the subject most likely to come up and then regurgitating the contents of her overworked brain. She related the difficulty of her subjects to Hans blow by blow, with comments such as:

> Everything has gone wrong today. The Latin exam was awful and I was so lost in contemplation of the paper, wondering how they dared to set just the pieces I didn't know, that I didn't notice what I was doing with my

hands. When I looked away from the paper I noticed that ink was everywhere – my hands, frock, handkerchief, exam paper – oh, it was a mess. My pen had been leaking evidently and I was playing about with it not noticing. My frock is quite spoilt – the brown one.

One can see how obsessively dependent she was upon her Swiss lover when she closed with, 'I'm realising that its very bad for me to rely so much on you . . . Whenever anything goes wrong I always want to dash to you . . .' Her fears on this occasion were groundless for within a few weeks she learned that she had been awarded a second class degree in Economics.

At last, she was Miss Johnson, BA.

Chapter four

PERCHANCE TO DREAM

I t did not take Amy long to realise that she had very little to offer a prospective employer now that she had decided that teaching was not for her, since apart from a piece of paper proving that she held a degree in Economics, she was completely lacking in any business experience.

Will Johnson had departed for America on a business trip at the end of May 1925 leaving his wife and two younger daughters, Mollie and Betty, to spend the next three months holidaying at Bridlington, but not before agreeing to finance Amy on a month's course at the Woods' Shorthand and Secretarial College in Hull. However, the timing of her attendance at the college did pose the question of where she would live whilst the family were away.

Amy knew without asking that she would not get the key to the family home, for her mother was far too astute to agree to leave her two older daughters with the run of the house whilst she was away. The only alternative, as Amy saw it, was for her to share lodgings with Irene on the Anlaby Road, an altogether unsatisfactory solution as far as her mother was concerned, but she was forced to accept it as the lesser of two evils. Ciss Johnson's mistrust might have been confirmed had she read the letter Amy wrote to Hans about it, in which she told him, 'Irene's arranged for me to go to her digs. We'll be there together – we shall have some fun.'

Finding a suitable secretarial job was even more difficult than Amy had imagined. She soon discovered that far from enhancing her prospects of employment, her degree became a hindrance, since in the eyes of those who

*This corner property in Springbank, Hull, once housed F.C. Wood's Shorthand &
Secretarial College where Amy honed her shorthand and typing skills in June 1925.*
(Chris Ketchell)

read her letter of application she was over-qualified. Employers were looking
for younger experienced staff, since they would have to start at the bottom
of the ladder and very often be required to perform the most menial of
tasks. Stamp-licking, filing memos and tea-making were not things that
graduates easily accepted.* Eventually, she found employment as a shorthand
typist in a chartered accountant's office in Bowlalley Lane at the meagre
salary of £1 per week, (at today's value just £100).

Amy hated the job right from the start and when she was asked to take
down shorthand for the Principal, Frank Hall, she found herself woefully
inadequate. Time and again she would have to go back to him, seeking
clarification on gaps in her notes. To add to the embarrassment, she was
constantly making typing errors, which meant retyping since she was not
allowed to make corrections. On the rare occasions when she was tempted
to make judicious use of the rubber on minor alterations, she was always
found out. How much easier it would have been with the modern word

* Gosschalks, a well known firm of solicitors in Hull, displayed a framed letter of a job
application from Amy in their reception office for many years after she became famous.

processor. To make matters worse, she found little support from the other girls in the office, who seemed to resent her as someone 'different'. Whether that was because she had a degree and was therefore regarded as someone to be taken down a peg, or whether it was because of a superior attitude on her part; one can only surmise. The only advantage about the job was that her office was situated less than 200 yards from the offices of Arregger & Raymonde. One can imagine the tales of woe Hans was made to endure.

It is not surprising that after less than three months of what was for her a highly stressful job, Amy suffered a nervous breakdown. No doubt it caused her parents some alarm, for Irene had gone through a similar phase with quite frightening results. After a family discussion it was decided that she should go to Bournemouth to recuperate. Amy imagined that she might be sent to one of the more salubrious hydros that existed in that well-heeled watering hole, and was quite disappointed when she found that she was to stay with one of her father's cousins.

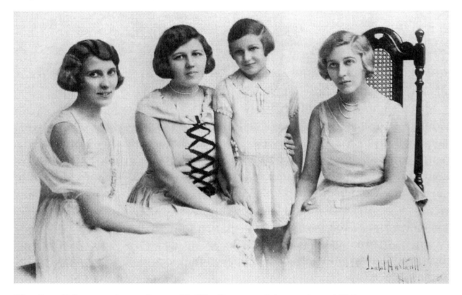

The four Johnson sisters, Irene, Mollie, Betty and Amy, circa 1925.
(The Amy Johnson Estate)

The Eddisons lived in a large detached house situated on the outskirts of the town, just north of the main railway line to London, a mile and a half from the seafront. Amy arrived in early October with strict instructions from her mother – reinforced by a letter to her hostess – that she was not to go 'gadding' to dances and cinemas, but that she was to have plenty of sleep, good food and rest. Her first impressions, as related in her letters to Hans, were that Bournemouth was a 'topping' place in which to live, with its magnificent houses set amongst pine woods, and its 'jolly fine gardens' with roses in abundance, so different from what she considered to be dreary old Hull.

There seemed to be only one minor flaw in Amy's new-found tranquillity, which came in the form of a five-year-old named Bunty, the Eddison's loquacious and inquisitive youngest child. She took an immediate interest in the family's new arrival, and Amy found that she was followed everywhere she went. Moreover, much to her annoyance, she was expected to take the little girl with her whenever she did go out. At first it was a love-hate relationship, but Bunty gradually wormed her way into Amy's heart with her perceptive questioning. On one occasion she watched Amy sewing and asked what it was she was doing and who it was for. When told that it was a night-dress case for her mother, she paused for a while and then said, 'Well, you are a nice old thing, aren't you?' Gradually, the two of them became the best of friends.

During the six weeks that Amy spent with the Eddisons she wrote twenty-six letters to Hans, sometimes as frequently as two in a day, so much so that Mrs Eddison was constantly reminding her young guest that she was wasting all of her valuable 'holiday time' writing letters.

Amy's sights were still set on getting Hans into a commitment which would lead to marriage. She prodded him with, 'Darling, don't you ever have the ambition to live in a beautiful house . . . I'm not beautiful enough, but p'raps I'd do if I'd a very pretty frock.' She desperately wanted a husband, a home of her own and children; such were the limited horizons set before most young women of that period. She craved affection from him with pleas of, 'I do love you, darling, and I adore you when you call me nice names.' She reminded him of the promise he made that he would take her to meet his parents: 'Are we really, honestly, truly going to Switzerland? Both of us together?' Hans continually stonewalled and at one point he scolded her for her insistence.*

Mrs Eddison appears to have been every bit as talkative as her youngest offspring, with her main topic of conversation being houses and their furnishing, since she and her husband were contemplating a move to a more expensive area. As a consequence, Amy found that she was required to accompany them, along with Bunty, whilst they toured the district house-

---

* Could it be that Amy was currently reading Ibsen's *Ghosts*, and that she saw herself as the frustrated Regina reminding the ill-fated Oswald that he had promised to take her to Paris?

hunting. It meant sitting in their car waiting outside properties for long periods as her hosts discussed some finer points of detail with the builders, or whilst Mr Eddison strode around the plot measuring up.

Amy soon avoided these frustrating excursions by asking for her and Bunty to be dropped off at the beach along the way and then picked up later when her hosts made their way back home. It was on such an occasion that she showed her dry sense of humour and keen eye for observing those around her. She was sitting quietly on the beach whilst Bunty amused herself on the sands, when another rather noisy family approached and sat down a few feet away. In a letter to Hans she wrote:

> It's simply lovely down here – blue sea, blue sky, sunshine, golden sands, bright yellow cliffs and up to a few moments ago – peace. Now it's 'Jeremy, come and put your hat on. Jeremy, where's the towel?' – then an awful rave to find the towel. Mamma shrieks into the air for Cissie, who presumably is on the top of the cliff (which is twice the height of Bempton Cliffs). It's a huge family of fat women, nursemaids and children, the most tiresome of whom seems to be Jeremy. There's miles of sand and they've planted themselves about three feet from us. I wonder if it would look very noticeable if we flitted. Jeremy's howling now.

The range of subjects which Amy covered in her correspondence with her lover was endless, and varied from her struggle with her debts – she was ever to be a hopeless spendthrift – to her health problems, which even included comments on the irregularity and severity of her monthly cycle, or as she would euphemistically term it, 'the wretched habit'. They would exchange views on the various authors that they were reading, and one cannot help but wonder if Hans was deliberately encouraging her to read the works of Ibsen and those of similar libertine views in order to make her more pliable sexually.

Although one knows from the content of Amy's letters that Hans was playing hard to get, there is no doubt that he enjoyed the flattery of her attentions and never hesitated to reprove her if she failed to write to him often. After being away a few weeks she wrote, 'You do seem to get fed up when there isn't a letter when you expect one, don't you? You're very greedy indeed wanting one every day.' She then lashed out at him in a feisty mood with, 'I shall write when I feel like it and have the time.' But even when angry with him she never slammed the door quite shut and always ended her letters with 'Beaucoup d'amour' or 'Your mon amour'.

Her mood swings, which she readily admitted to, and which might have been caused by some extreme form of premenstrual tension, were quite evident in her letters; she could be angry and abusive at times and then not post them. One presumes that for her there was some form of catharsis in writing them. She would then explain to Hans why she had not sent them, whereupon he would be curious and ask to see them. Surprisingly, she would send them. It was all part of the psychological warfare which seemed to be endemic in their relationship.

Apart from their disapproval of Amy smoking cigarettes in front of Bunty, the Eddisons appear to have had a favourable impression of their convalescent guest. She was quick to remind Hans of this, and of the fact that Mr Eddison had 'taken quite a fancy' to her and had invited her to stay on over Christmas. She obviously hoped that Hans would implore her not to stay, but in fact she had already declined as graciously as she could, for she was not only dying to see him again but also wanted to be present at Irene's twenty-first birthday party at the end of November. It had always been a sore point with Amy that she had not been given a similar party or dance on her own coming of age. However, her affable father sought to make up for this by sending her 5 guineas, so that both sisters could have new dance frocks for the occasion.

It was decided that the dance would be held in either the Metropole or the El Negresco, two of Hull's more salubrious hotels. However, it presented more than a few problems for both sisters and parents, since they all had different ideas of how it should be conducted. Irene wanted to invite forty guests of her own choosing, reminding Amy she was short of girls and not boys! She also wanted her birthday toast to be with champagne, something her parents strongly resisted. Matters became even more gloomy for the two sisters when their mother insisted that a large number of family relatives be invited. The prospect of her mother's wishes being fulfilled caused Amy to describe to Hans Irene's vision of 'the ballroom floor covered with aged, frumpy couples trying to dance, as most distressing'. As a last resort, Irene planned to hold a whist drive in a separate room adjacent to the ballroom and hope that the older relatives gathered there!

Like many of their generation, Amy and Irene were captivated by the modern dance music and the foot-tapping syncopated tunes of that era, and at home they would don their headphones and tune in to the wireless transmissions from the BBC's 2LO station at Savoy Hill. The British Broadcasting Company was founded with Post Office approval in 1922, and it was not long before modern dance music was being heard over the airwaves from bands such as the Savoy Orpheans and the Savoy Havana Orchestra. The American influence was very much in evidence, from the lilting piano-playing of Carroll Gibbons, to the 'voh-voh-de-voh' crooning of

Rudy Vallee with his hand-held megaphone. The latter was a style of music which soon had a cult following amongst the young.

Amy told Hans how much she appreciated the fact that the Eddisons had a most modern innovation coupled to their wireless – a loudspeaker – which meant that she could dance freely with Bunty as her partner without the restriction of headphones. One of the popular songs at the time was 'Tea for Two', and listening to the lyric today one can easily understand how it fitted Amy's mood during that stage of her affair with Hans. In a postscript to one of her letters she reminded him that she would not be happy until she was 'sitting on his knee with his arms around her'. One wonders just how she resisted the temptation of adding, 'We shall raise a family, a boy for you, a girl for me', since the lines must surely have been running through her mind at the time.

During the twenties the tango was all the rage and Amy asked Hans if he could arrange for her to be taught the new dance as soon as she arrived home, adding, 'I'd hate it if Irene danced it and I couldn't.' Amy always suspected that Hans was attracted to her sister and in what must have been one of her most vitriolic letters to him, she lambasted him mercilessly with, 'Have just got a letter from Irene where she tells me she met you at our house and you took her home and she had to listen patiently whilst you talked about *me* all the way home. How dare you discuss me with Irene – I am furious about that. How would you like it if I met your brother and discussed you with him!' There is no doubt that Amy had allowed her imagination to run riot on this occasion, because Irene was currently being courted by a young man whom she and the family had known for some considerable time.

Teddy Pocock. (Marjorie Pocock)

Irene's aspirations for a stage career had led her to team up with a talented pianist and singer by the name of Edward Pocock, and she would accompany him with a coquettish song and dance routine when they entertained as an amateur duo. Teddy, as he was affectionately known to friends, was a well-mannered young man in his early thirties and came from a musical family who lived on the Holderness Road. His father sang with a local opera company and it seems most probable that Irene came to know Teddy through the musical interests the Pococks shared with Ciss Johnson.

From Amy's correspondence it seems that Hans would often tease Amy that she did not love him as ardently as Irene loved Teddy. This type of comment would annoy her intensely, to the point where she would reply, 'You all stand enslaved by Irene's vivacity and charm', and then add, 'She is the sole subject of conversation at home. I do get hatefully jealous of her and yet I'm awfully fond of her.

There were two rather ominous accounts in Amy's letters whilst she was at Bournemouth, both of which had a prophetic ring about them. The first occurred in one of her earlier letters, in which she asked Hans if he believed in the interpretation of dreams. She then reminded him of a dream that she had told him about previously, in which they were not able to be married because she could not find her white gloves. She then asked him if he thought the dream had any significance, adding that she did.

In the second, she told Hans of a conversation with Mr Eddison, when he warned her of the hazards of the sea. He told her how he had once been invited to go fishing with some friends, but was forced to decline because of an appointment with his dentist. The boat capsized and his friends were never seen again! He then proceeded to describe in gruesome detail the sight of a man's body which was washed ashore at Swanage. Lapping about in the clear water the badly decomposed body was partially eaten away by lobsters, which were still clinging to it. The story was one which Amy could never quite forget.

There was great excitement in the Eddison's household on the Thursday afternoon of the last week in October 1925, for as the family were just about to sit down to tea they spotted the postman walking up the front path with a telegram in his hand. Their two elder daughters were suffering from chicken-pox and were in quarantine at their boarding school in Swanage and as their mother hurried to answer the doorbell, her immediate thoughts were that it was to say that they would not be allowed to come home for the half-term holiday. Her relief was mingled with surprise when she saw that the telegram was addressed to their guest. Amy tore the yellow envelope open to read that Hans was travelling to London on business on the next day and that he was prepared to combine the trip with a visit to Bournemouth for the weekend, if that was convenient. The Eddisons were quite in agreement that Amy's boyfriend should stay and a telegram was sent back telling him that he was welcome.

Evidently Hans made quite an impact on the family for in a subsequent letter Amy was pleased to tell him, 'Mr Eddison says I'm very silly to go to America and risk losing you . . . and that he'd bet me £1000 you wouldn't still be a bachelor in another twelve months. You've made a great hit here.' Even Bunty was smitten, for she wrote a letter to Hans, but when Amy asked to see it the little girl coyly replied, 'No, Mammy might not like it', and then proceeded to throw it into the fire.

Much to Amy's disappointment it was a wet weekend when Hans stayed with her, but it did not prevent her from showing him around the town, for at that time she made it known that Bournemouth suited her 'splendidly'. She loved its mild climate, the smell of the pine trees in the wooded chines, the abundance of landscaped gardens and the cliff-top walks which she found so invigorating. Her dream would have been to marry Hans and settle down right there and then, but for the time being she had to be satisfied with this transitory visit.

They walked and discussed the books they had been reading, the author of the moment being Henrik Ibsen, for ever since Amy had attended a lecture and made notes on the man and his works she had been anxious to read his plays, explaining 'he appeals to me very strongly – he's wonderful'. Soon after Hans returned to Hull, she thanked him 'awfully' for traipsing around the bookshops in the pouring rain trying to buy a copy of *Brand*, and then suggested that they read it together when she returned home. It was all a world away from the time when she had been a Sunday-school teacher at her father's chapel in St George's Road.

The highlight of the weekend for Amy had been their visit to the Royal Bath Hotel on the Saturday evening to attend a dance. There is very little possibility that Amy let Hans out of her sight on the dance floor on that evening, since she knew from the time he visited her at university that he liked to flirt with the girls. On that occasion she had reproved him for his over-familiarity with some of his student dancing partners. It was an incident that she had not forgotten and whenever he told her that he would be attending a dance, whether in Hull or Liverpool, she would remind him with comments such as, 'I'm longing for you to put your arms around *me*. I hate to think you would go and have another partner. You might be induced to kiss them like you did at Sheffield.'

Whilst there is a fine shade of distinction between a womaniser and a 'ladies man', Amy was quite sure that Hans could easily qualify as the latter. She was never quite certain where she stood in her relationship with him, for he could be unaccountably moody and he was often more ready to criticise her than to praise her. His letters could be 'nice ones' as she termed them, or they could be disquietingly cold and formal. How much of this formality and distance was due to his being a foreigner and therefore unable to express himself as much as he would like to in an unfamiliar language, one can only guess.

Any coldness that Hans might have displayed that weekend would not have been due to any lack of attention from Amy. Although the Eddisons were accommodating a larger than usual complement of the family that weekend, in what Amy described as 'a bit of a squash', it did not deter her from tip-toeing across the hallway at five in the morning and entering her lover's bedroom!

Amy lived on her memories for the following two weeks, telling Hans that she had never felt so fit and full of energy and that, a few days before she left for home, she 'ran down the zig-zag path again to the sea, startling the people on seats at the corner'. Here was a woman deliriously in love, and at the same time soaking up the images of her stay in what was for her a sylvan idyll, taking it all in but not knowing where her life would lead.

One wonders if any passer-by noticed the diminutive form of the young woman and her tiny friend skipping along beside her as they made their way back for the last time through Meyrick Park. How were they to know – how was *she* to know – that in less than five years' time those green lawns would echo to thousands of people chanting 'Amy, Amy, Amy!'

Chapter five

THE HESITANT SWISS

my's first task upon arriving home from Bournemouth was to find new employment in Hull, now that she had decided that she would no longer go back into Frank Hall's accountancy business. Her two other immediate concerns were to become financially independent of her parents and to clear what was for her a heavy debt of some £33 (an amount which might appear ridiculously small, but which in today's values would amount to approximately £3000). It was a sum which she had incurred by her extravagant spending on clothes during her time at Sheffield and whilst living with the Eddisons. Stockings and dance frocks appear to have formed the bulk of her debt, and she was apt to be over-extravagant when buying gifts for her three sisters. Hans had come to her rescue, but true to the tradition of the Swiss businessman he aspired to be, it was not without strings attached. Amy gave him some of her War Savings Certificates as a deposit, pointing out that whilst they were immediately cashable, they did not fully mature until 1928. To be fair to Hans it might well have been that she offered them rather than that he asked for them.

In her memoirs, *Myself When Young*, Amy described how she was flicking through the pages of a glossy magazine one day and taking particular notice of the design of the advertisements. She found herself criticising the wording and layout of some of them, and partly for amusement she started to doodle and revamp them with designs of her own. Thus was born the germ of the idea that she would aim for employment in the world of advertising. Scouring the local newspapers she saw an advertisement for staff vacancies by Morison's Advertising Agency in Albion Street, quite near to where she had previously worked. She recalled:

> I did not ask for a job but the opportunity to learn the business, and I
> secured an interview with a leading agency. Here I found that one of the
> partners needed a secretary, and I was able to make an arrangement
> whereby I was to give my services as secretary in exchange for learning
> the art of writing advertising 'copy' and planning advertising campaigns.
> In addition, I was to receive a nominal salary of thirty shillings a week.
> This seemed an admirable arrangement and I started work very soon
> afterwards, full of energy and enthusiasm.

Once she had landed the job with Morison's, which was close to where her
lover worked, her involvement with him deepened, and her parents became
alarmed at the prospect of their daughter marrying Hans Arregger. The
knowledge that Amy was his constant companion at weekends, riding on the
back of what was to them a dangerous vehicle – a motorcycle – was enough
to exacerbate her mother's 'nerves' in the first place, but it was nothing
compared with the fact that their potential son-in-law was a practising
Roman Catholic *and* a foreigner to boot. That scenario would have been
completely unacceptable to Ciss Johnson, who was a staunch believer of the
Methodist persuasion.

Although Hans was a frequent visitor to 85 Park Avenue, where he seemed
to get along reasonably well with Amy's father, he was never quite the
persona grata that Amy would have wished him to be. Mollie and Betty, like
most younger sisters, viewed their elder sister's romance with a studied
curiosity. Betty nicknamed Amy's boyfriend 'Pansy', not so much because of
any effeminate connotations, but merely because his name rhymed neatly
with 'Hansy'. Meanwhile, Irene was far too engrossed with her own 'dear
Teddy' and their impending engagement to take much notice of her sister's
affair.

Amy became quite adept at covering up the secret weekends that she
spent with Hans, excusing her absence from the family home by telling her
parents that she was visiting one of her old university friends, and so it was
that the couple holidayed together somewhere near Loch Lomond during
Easter 1926. In a letter written to him whilst he was in Hamburg on business
just after their time together in Scotland, she reminded him of their visit and
added, 'You were such a dear all the time on our "honeymoon" and I don't
think I ever was so thoroughly happy. I can remember every detail, can't
you? Hull is so empty when you're gone.' She then signed off with '*Ta chérie*
(once your wife)'.

Although Hans had given Amy assurances on several occasions that they would be married 'some time', she was not slow in reminding him that this was not good enough, and in her subsequent letters she became much more demanding. One can sense the steely resolve that was to show itself in a quite different way in later years. She accused him of being 'as close as an oyster', and berated him with 'You are quick to be angry if I assume there isn't any proprietary rights over me.'

It becomes transparently clear that there was a great deal of hypocrisy in Hans's stalling tactics. She reminded him that whilst he wanted her to become an emancipated woman, throwing off her inhibitions by flouting the social conventions and restraints of the day, that at the same time he claimed rights of ownership over her. It was all a far cry from Ibsen's espousal of liberation for women as portrayed in *The Doll's House*.

One can easily blame Amy for her weakness in giving in to him, and she freely admitted to it. However, she continued to play along with him in the belief that he would eventually marry her. This hope was heightened significantly when Hans finally agreed that she should accompany him to Switzerland to meet his family.

In mid-June the couple sailed from Hull to Zeebrugge and boarded a train bound for Lucerne, where an excited Amy was welcomed as Hans's 'new girlfriend' by his widowed mother and three sisters, Emily, Hedy and Mary (a fourth sister Fridy was living in Paris). From a photograph taken at the time, one can see Hans, his mother and one of his sisters, with a smiling Amy, seated around a table having drinks together on the balcony of the family home in Hochbuhlstrasse. One cannot fail to notice the happiness radiating from Amy's face as she and her lover, his hand on hers, turn to face the camera.

There is no doubt that Amy fully expected Hans to announce their engagement during the fortnight that she was in Switzerland with his family, or at least make his intentions clear to her. Sadly, she was disappointed. One suspects that Hans's mother, a strict Roman Catholic, would want to impress upon her son the necessity of Amy converting to his faith if he were to marry her, although this does not appear to have surfaced immediately in letters written after her visit.

Amy's expectations reached their peak on 1 July when a party was given to celebrate her twenty-third birthday. The whole family gathered together with some of Hans's friends in one of Lucerne's most famous restaurants, *The Wilden Mann*, on Bahnhofstrasse. As the meal proceeded, the wine flowed more freely and inhibitions were lowered, causing Amy, who was a teetotaller, to become slightly uneasy as she watched her lover flirting with some of the girls around the table. To make matters worse, they were linking arms and kissing and conversing in a language which she barely understood. There is no doubt that Amy felt humiliated by the insensitivity of Hans's behaviour.

Before the party finally broke up, Amy was given wedding presents by two of Hans's friends. Dr and Mrs Egli gave her a blue vase, whilst another friend, Hans Vetter, presented her with two pictures of local Swiss scenes, gifts which he presumed would be for the couple's future home. It was a very emotional moment for Amy, and during the ensuing conversation, whilst she was busily thanking them for their generosity, Hans slipped away with one of his friends, quite unnoticed. Left on her own and feeling more than a little embarrassed, Amy decided to visit the Roman Catholic cathedral nearby before going back to the Arreggers' home. She knelt and prayed that Hans would not break his promise.

Amy always found it much easier when she had something of deep importance to say to Hans to express her thoughts on paper rather than tackling him face to face, and in a letter written not long after she had returned home from Switzerland, she launched into one of her bitterest attacks over his behaviour towards her:

> Dear Hans
> I can't let you go out of my life for always without saying a few things to relieve my mind a bit, as I can't tell you personally I'd better write them and then there can be no misunderstanding.
>
> Now at last I understand what you meant once when you said you were afraid I had illusions about you which would some time be shattered. You know that I have loved you almost ever since I knew you nearly five years ago. You know that I have practically worshipped you. You know that I have offered you everything . . . my career, my ambitions – my whole life. And you not only didn't want what I wanted to give you, you despised it. I couldn't give you enough. It didn't satisfy you – you found fault with everything I was giving you.
>
> This weekend we were away was a further proof to me. You were absolutely indifferent to me. There was no romance left to you in us being together. Why do you think I hated to put that sixpenny brass wedding ring on? Because it was typical of our relationship together. Just make-believe, not the real thing – and not even a beautiful romantic substitute for the real thing, but just a sordid, common, matter-of-fact invitation. I hate that ring – I couldn't wear it again. It was different in Scotland – you loved me then. I was quite happy. How quickly you have tired of the novelty of having me to yourself.
>
> You used to love my hair, my fingers, my eyes, my legs, my body – do you ever think of them now? The only remarks I ever hear from you now about my personal appearance are: 'You've got a mark on your frock.' 'That bow looks like a bag.' 'Your hair's like a golliwog's.' 'Your arms are

getting thin.' 'What stockings are *those?*' We haven't been married for thirty years by any chance, have we?

Switzerland was the first big blow to my faith in you. It was a very big blow to me. I had a premonition that you wouldn't keep to what you had promised, and as the time got nearer to my birthday I knew you intended to forget. Even right up to the moment on the stairs when I asked you about it, I made myself believe that you had merely forgotten, and I told myself I was a fool to regard it so seriously, but at that moment I knew that you had forgotten on purpose. Oh, it's terrible to see my most cherished dreams being regarded just as a joke by you, whom I have adored. Of course you are highly amused by this so far, and just a little irritated. You think I am melodramatic – maybe so – I don't know. Why did you think I wanted to smash the vase *your* friends gave to both of us. I love it and I hate it.

You haven't broken my heart, but you have made me very bitter. I know you will find this letter just as funny as the rest of my most intimate feelings. When I tell you you've hurt me, you laugh. But I think some day you will have something to answer for. You say you love me, but what is your love worth when you let me go away for always without an effort to stop me [she was still threatening to go to America]. I still love you, and I can't believe that all I have written is true. If only you could tell me it isn't.
Yours,
Amy.

One finds it incredibly hard to conceive just how Amy could have held on to such a tenuous relationship, since Hans appears to have toyed with her emotions almost as trivially as a kitten might play with a ball of wool. Her love for him was so palpably intense and loyal, and yet even when she saw that it was not reciprocated, she still clung to the belief that matters would improve.

There is no doubt that he knew he had complete power over her emotionally, a power which he did not hesitate to exploit to the full. Moreover, she made it easy for him to control the relationship when she ended her tirades by continually leaving the door of opportunity slightly ajar – not too much, but just enough to salvage her pride. She is never quite prepared to end the affair. He, on his part, was quite content to keep the status quo by going along with the role that she was pursuing. He placated her with gifts and endearments which she eagerly accepted, and it is therefore no surprise that we find her writing to him later in the year, whilst he was in Germany, fully reconciled to playing a waiting game.

In order to prevent her parents from knowing how deeply involved she was with Hans, particularly after she returned from Switzerland, she asked him, whilst he was abroad on business trips, to send his letters to her at the office rather than direct them to her home. It was not an ideal situation, for his letters would arrive at Morison's with all the other foreign mail and there was therefore the chance that they could be opened before she received them, but it appears it was a risk she was prepared to take.

Amy had been slowly paying off the debt she owed Hans week by week, leaving her little to spend on herself after she had paid her mother for her keep. This made her all the more determined to ask her boss at Morison's Advertising Agency for a rise, but on each occasion, over a period of several months, she was fobbed off with the excuse that the firm had barely enough to pay staff salaries, let alone give increases. Their refusal was like salt in a raw wound, for she was already being asked to perform menial tasks well below her ability, such as folding circulars and putting them into envelopes. Not only was it a task normally given to an office junior, but it also meant she had to work late in the evenings and on some occasions until 8.45 p.m. without any extra recompense.

It is worth noting that it was quite common in the 1920s for office workers to leave home at 6.30 a.m. and not return until 7 p.m. Staff were expected to work until the job was done, which meant putting in many hours over and above their normal ones, without payment for overtime. In addition, they would be under vigilant supervision imposed by a rigid pecking order of authority. Such frustrations, together with the growing opposition from her parents, finally caused Amy to question whether she should stay with Morison's. Knowing that Hans made regular business trips to London, she now began to consider working there. No doubt the prospect held out the added attraction of being able to spend weekends with her lover without her family's knowledge – or so she thought.

Much has been made of the fact that Amy took up flying as a direct result of a 5-shilling flight she had with Mollie on a Sunday afternoon early in November 1926. One would like to think that this was so, but there is very little evidence to support it. It appears that it made very little impression on her, other than that, as she told Hans, 'we both enjoyed it, but I would have liked to have done some stunts'.

Flying was a novelty in the twenties and the Surrey Flying Services of Croydon were touring the country giving rides in one of their Avro biplanes. When they came to Humberside they used a farmer's field just off Endike Lane, not far from Amy's Aunt Campey's home at Cottingham.

As soon as news of their visit came to the notice of Amy and Mollie, they cycled to the field and queued patiently with many other would-be adventurers. When it came to their turn to fly, they climbed a small wooden stepladder to access the open rear cockpit, which accommodated just four passengers sitting side by side in pairs. To assist these underpowered machines into the air with five on board, on what were merely short-hop flights, the pilot would usually keep the fuel tank as low as safety permitted. Passengers were rarely in the air for more than five minutes, and the whole affair came across as rather tame to the Johnson sisters. But what did it matter? It at least gave them the kudos of being able to tell of their experience to family and friends later on.

Thrill of a lifetime for five shillings! Passengers embark into a five seater Avro 536 at Canvey Island, Essex, in May 1923. Amy and Mollie made their first flight in a similar machine. (Hulton Getty Picture Collection Ltd.)

Irene's engagement to Teddy in 1926 made Amy all the more determined to lead Hans in the same direction. She was not slow in reminding him of the

Irene Johnson at the time of her engagement to Teddy Pocock. (The Amy Johnson Estate)

fact that her sister's courtship had been a brief one and that she and her boyfriend were now preparing for their wedding in the summer of the following year.

From an account in one of Amy's letters to Hans, where she interrupted her train of thought to write, 'Oh, bother, Irene and Teddy have come in and they're squabbling awfully – I've sent them in the other room to finish it', one wonders if her sister and future brother-in-law were really compatible. But perhaps it was nothing more than a lover's tiff about furnishings for their new home, now that they had decided to purchase the house next door to Teddy's parents.

Employment prospects were at an all-time low ever since the General Strike, which had threatened to paralyse the whole nation, but both Teddy and Irene were largely unaffected by it. He had a steady job in the local offices of Shell Mex, whilst she was about to start as the manageress of an advertising department in a large local store in Hull at the end of November. However, combining Amy's snippets of gossip about their quarrelling with the knowledge of Irene's wild and erratic behaviour in the past, one wonders if these pointers were a premonition of their doomed marriage.

In December 1926 Hans travelled to Lucerne to spend the Christmas and New Year with his mother and sisters, a trip which he combined with business that he had in Italy. In a letter written on the day after he left, Amy told him that she would be thinking of him especially on Christmas morning whilst he was in church with his family, adding that she hoped to be with him in spirit, if not in person, by worshipping at the same time. And so it transpired that Amy's relationship with her parents reached their lowest ebb, when they discovered that she had slipped out of the house early on the Christmas morning to attend mass at St Vincent's, a Roman Catholic Church nearby.

This must have been the final straw for Will and Ciss Johnson, and from here on things became unbearable for Amy. There is no doubt that over the next few weeks her parents must have given her some form of ultimatum. Hard decisions were made. Either she must give up the idea of marrying this man, a papist and a foreigner, or she must leave.

Chapter Six

'HELL HATH NO FURY...'

ans had already decided before leaving for Lucerne that he would move his lodgings to more desirable quarters when he returned. It was a decision which pleased Amy for she had always disliked the drab room he rented and in particular his manservant, Williams, who seemed to disapprove of Amy's evening visits, when she stayed late.

Amy was determined that she was not going to be obstructed when Hans moved, and she decided to call on his new landlady on the pretext that she was his sister. She visited the house and dropped the hint that she would be visiting him regularly when he finally took up residence. Whether the woman believed her or not is questionable; the more shrewd members of her profession would have known that such tricks were not uncommon amongst lodgers who wished to entertain guests of the opposite sex in their rooms. Having trodden on such dangerous ground, Amy had misgivings about telling Hans what she had done, fearing that he would be angry with her. However, it appears that he regarded the incident with little more than mild amusement and was, if anything, rather flattered by the zeal with which she pursued him.

As it happened, Amy had little chance to avail herself of her carefully laid plan for clandestine meetings with her lover, for within a few weeks, soon after the New Year, she received an offer of employment from one of the agencies in London to which she had applied. The John Lewis store in Oxford Street was prepared to offer her a learnership post in their subsidiary store, Peter Jones Ltd, in Sloane Square. Although on the surface it looked to be a promising opportunity, it had certain drawbacks which were

Amy in the garden of 85 Park Avenue, Hull, not long before she left home to work in London.
(The Amy Johnson Estate)

not apparent to Amy at the time. Much as her parents wanted her to break with Hans, they were dismayed when they learned that she was about to move away to London. However, she accepted the offer and booked a room at the YWCA in the Ames House Hostel at Euston, until she could find more permanent accommodation.

During the first weekend in March the two lovers travelled, as she put it, 'to town' (an affected term which she liked to use for a visit to London), where it appears that they stayed together in one of the capital's hotels. It might well have been at the Strand Palace Hotel, for in a letter written a few days later, she told Hans that after seeing him off at Euston, she had gone back to this hotel and retrieved his gloves from under the table where they had been sitting.

On that Sunday they attended the morning service at the Roman Catholic Westminster Cathedral. It left them still grappling with their unresolved religious differences, and Amy seems to have been surprised herself at how deeply ingrained were her own Nonconformist beliefs, for she wrote, 'I was entirely in sympathy with your wish to bring up any children of yours in the Catholic religion . . . It was rather unfortunate that the sermon just emphasised those very points which I had found so very difficult to face in that religion.'

Life in the YWCA hostel after the comparative freedom and luxury she had enjoyed at home was worse than Amy had imagined, and she described the 'wretched room' she was occupying to Hans in detail. It was only made bearable by having his photograph on her dressing-table, and many of the small gifts that he had given her over the past five years. She mentioned how she was awakened at 7 o'clock every morning with alarm bells of bed-shaking intensity, and how lights were turned off at 10.30 p.m., so that she was forced to write by the glimmer of a nightlamp and two candles when she came in late, as she often did.

Amy's sole consolation during the homesickness of that first week away was the hope that Hans would soon join her permanently in London. She comforted herself by telling him, 'I have made myself feel happy again with thinking about your coming to live here with me. It would be cruel if you let me go on thinking and planning and then tell me when you come that we can't be together.' She then added in an almost prophetic vein, 'You have given me so much and done so much for me, and although I would do anything for you, and give you everything I have, it seems so little to repay you. But perhaps some day I'll be able to do something really big for you, to show how much I love you.' She signed off 'Amy (or Mrs Mary Arregger)', showing not only the status for which she longed, but also how, at that time, she loathed her first name.

Amy was not one to let an opportunity slip and even before she started at Peter Jones, she had applied and been interviewed for a post which was being offered in the advertising department of Harrods store. Sensing that she was relatively inexperienced in this field, she was asked to write some sample copy and submit it, whereupon she would be given a decision.

It was during this period that she met up with Olive Birkbeck, a young woman who was to play a large part in Amy's economic survival during her first year in London. Olive and her sister Clarice were the stepdaughters of Amy's Uncle Alfred Hodge – Ciss Johnson's brother – who had married a widow living in Leeds. Olive was a staunch admirer of her clever university step-cousin from Hull, and much to Amy's embarrassment, she would sing her relative's praises with a great deal of embellishment whenever the opportunity arose. She was employed in the stationery department of the Times Book Club and had advised Amy to apply for a post in the company's library. Meanwhile, Amy's latent adventurous streak caused her to cast an eye on the possibility of a job on the continent, whereupon Hans advised her to apply to a Swiss employment bureau in the City.

Nine new members of the learnership scheme, all 'gentlemen and ladies of good breeding', started at Peter Jones during the second week of March on a four-week probationary period. Amy, wearing an outfit supplied by the company, consisting of a smart fawn frock with matching silk collar, was designated to assist in the silks department. Initially, it was quite to her liking, for it reminded her of playing shops when she was a small child. In *Myself When Young*, she described that first Monday morning.

> I took my place behind the counter to sell chiffons, silks, satins and ribbons. I loved the delicate shining materials and was quite prepared to throw my whole energy and enthusiasm into making a success of this new job. Being the junior and last-comer of the department's five assistants, I was only allowed to serve when all the rest were already busy – which never happened. I spent the whole of every day on my feet from nine o'clock to six, unutterably bored and tired, for nothing is worse than inactivity. The only jobs found for me were tidying away the boxes rifled by senior assistants.

To make matters worse, she found that she was at the beck and call of senior staff and made to run to and fro throughout the store on what she deemed to be unnecessary errands, the only respite for her aching feet being in the form of a twenty-minute 'sit allowance' spread throughout the day. An added ignominy was the necessity to clock on and off at the start and finish of each day, a duty which irked Amy considerably, since the time spent queuing after 1 p.m. to clock off on a Saturday afternoon (several hundred were employed by the John Lewis organisation in London) ate into her precious leisure time. To add to the harshness of these conditions, holiday pay was non-existent, whilst sick pay was at the discretion of the company.

It took Amy exactly five days – until the first pay day – for the fact to sink in that she was being exploited. The company ran what it termed a 'remuneration account', which meant that although she was given an advance of £.3, she was required to submit a valuation of her services at the end of each week. More importantly, it was a sum that had to be agreed with, by her supervisor. Much to her dismay, she was valued at 8s 4d during her first week, which meant that she now owed the firm £2.11.8d! Commission could be earned if a learner recommended the name of a friend or acquaintance who might subsequently become a customer of the store, but, being a comparative stranger in London, this was of little value to Amy. In a letter to Hans she told him that she would be owing the firm a lot of money if she stayed; one girl had wanted to leave and found that she owed £60. And then with a deft touch of humour she added 'I'll have to be bought from the shop like a yard of silk.'

Apart from the prospect of failure at Peter Jones, to which she seems to have become resigned, she failed to obtain any of the three other posts she had applied for, at Harrods, the Times Book Club and the Swiss agency; the latter was looking for staff who were fluent in French or German. Matters were made worse by the grim conditions she experienced at the hostel, where she was having to boil water in a pint-sized kettle on a small gas ring, and then carry it up and down a flight of stairs to do her washing, the cost of laundry services being beyond her means. Pressing clothes necessitated the removal of the bulb from the one and only overhead electric socket and plugging an electric iron into it. This would have been no problem when there was daylight available, but for Amy it meant ironing during the evenings by candlelight. Even then, it had to be completed before the supply was cut off at 10.30. These conditions made her all the more determined to find a flat or bed-sit, if possible nearer to where she was working; or if she could obtain more suitable employment, one nearer to where she intended to work. It was Olive Birkbeck who came to her rescue on both counts.

Olive was the only friend that Amy had in the lonely capital in 1927, and she was kind enough to invite her home for weekends to stay with her at her brother and sister-in-law's home in Winchmore Hill. It was during one of these visits that she told Amy of a Yorkshire relative who would be prepared to take her as a lodger on quite reasonable terms. The average charge at that time, inclusive of meals, was 35 shillings, but Olive's relative, Mrs Green, was prepared to accept just 25 shillings until such time as Amy's salary improved. This was a big help to Amy, and as a result she moved out of the hostel after three weeks to live in West Hampstead. The only warning that Olive gave her was that Mrs Green's brother, a widower who lived in the downstairs flat, was 'a bit keen on the girls and needed keeping in his place'. Amy took due note of the advice and moved in.

During the last week of her four-week probationary period at Peter Jones, Amy developed influenza and decided not to return to the store. She was now uncertain about the debt she had incurred, for in the second week she had only been valued at 10 shillings and the debt was mounting. She needed expert legal advice, and again the ever helpful Olive knew exactly where she could find it.

The genial Vernon Wood was one of the senior partners in a well-known firm of London solicitors run by William Charles Crocker (later to be knighted), at 21 Bucklersbury, quite near to the Mansion House in the heart of the City. When Amy was first introduced to him, she did not realise that he had known her as a small child, for he was a friend of the family on her mother's side. He was also in regular contact with Olive, who felt that he would be able to advise Amy on the implications of her employment contract with Peter Jones, and how she could extricate herself from it.

The outcome was that the two women were invited to be Vernon Wood's guests at the Holborn restaurant for what Amy described as 'an awfully expensive meal'. The kindly solicitor's verdict was that she should leave the store and seek other employment. More importantly, and much to her surprise, he offered her a post as a typist at Crockers starting at £3 per week, with increments as and when she gained the necessary legal experience. It was an offer she gladly accepted.

In spite of the quarrel Amy had had at home over her relationship with Hans, her family never failed to be concerned for her welfare, particularly during those first few months whilst she was in London. Ciss Johnson visited her whilst she was at Peter Jones, waiting discreetly outside the store until her daughter finished work. She then took her back to the hotel where she was staying so that they could freshen up before going out for a meal with Olive. The rest of the evening was spent at the theatre. Amy relished her night out, telling Hans, 'I chose the wine – a half botle of 1914 Bordeaux – can't remember the name – but it was nice, not a bit sweet and not heavy.' She then added as if to pay homage to his tutelage as a gourmet, 'I was so glad I knew a bit about it.'

The correspondence between Amy and her lover continued unabated thoughout the first year that she was in London, the leitmotiv of her letters being 'I want you, I want you. When are you coming?' He continued to prevaricate with 'Let's wait and see', until in a burst of anger, she would suddenly accuse him of dishonesty in their relationship by quoting back to him a comment made to her by one of his Swiss friends, 'A charming boy but not the marrying kind.'

There is no doubt that Amy was on an emotional rollercoaster at this time. There were loud screams and white knuckles during the descents, and feelings of euphoria on the highs. Sometimes she wanted to have children, which inevitably meant that she would consider nothing less than marriage, and to this end she would seek to entice him into conjugality by enrolling for cookery classes to show her desire for domesticity, or German lessons so that she could be in touch with his culture. Then, in a quite different mood she would tear into him with, 'I don't care a damn whether you're a marrying man or not. Whatever difference does it make! I'm fed up with such talk and wish the institution of marriage was at the bottom of the sea.' What rankled with her most, was what she described as his 'dog in a manger' attitude. He did not want her to go out with other men, but at the same time he failed to make any firm commitment to her at all.

Studying Amy's letters, some seventy years after they were written, is much like attempting to assemble a jigsaw puzzle with many of the pieces missing, since we have none of Hans's letters to her. However, one can discern the man's character from the comments she made in reply to his letters. He appears to have been insensitive and impervious to her feelings, and there is no doubt that he considered himself to be superior to her. He treated her more or less as an affectionate but troublesome child, rather than a potential wife, and contributed to the lowering of her self-esteem by belittling her. In one letter she wrote, 'You appear to think it's [working in a shop] about the best profession I can take up.' Amy was not one to be walked all over and she hit back, saying, 'I'm impatient with your preaching. Couldn't you temper your lectures with just a little touch of humility?'

In spite of the psychological warfare being waged through the post, when his letters were considered to be either 'ripping' or 'nasty', they continued with their secret 'honeymoon' holidays and weekends together. At Easter, they spent four days in Wales, just after Amy had started at Crockers.

One gets the impression that Amy was never very happy with all the deceit that accompanied these assignations, and that she was troubled by deep feelings of guilt. Inevitably, she had to make up excuses for her parents, on this occasion telling them that she had met some friends who owned a car and would be motoring with them in Wales over Easter. In a letter to Hans just before the holiday, she wrote, 'How I do loathe all this lying and deception. I used to think it fun, but now it irritates me . . . Occasionally you say or do something which gives me courage to stick to my old ideals, which I am sure are right, though you scoff at them.' It was around this time that Amy had started going to church again, albeit sporadically, and most probably through the influence of her landlady, who appears to have been an earnest churchgoer.

The more Amy saw of Hans the more she wanted him. Both knew, but were unwilling to admit, that the main attraction between them was physical. She was unable to see what Olive Birkbeck and Winifred Irving could see clearly, that the relationship was going to end in tears. They knew that she was completely dependent upon a man in whom she confided every detail and facet of her life.

One cannot help but coming to the conclusion that Hans was so flattered by Amy's attention, even worship of him, that he could not let go of the relationship any more than she could. The very fact that her letters have survived, and that he kept them almost to the end of his life, poses an important question. Did he cherish the letters becase of his love for her, or were they merely a means of nourishing and gratifying an over-inflated ego?

Irene's marriage to Teddy had been planned for mid-April 1927, but was delayed until the last Saturday in May because the bride-to-be had been admitted to hospital a few weeks earlier with an infectious illness (most probably scarlet fever or diphtheria – common enough in those days, even amongst adults). There can be little doubt that Amy was invited to her sister's wedding, although there is no mention of it in her letters to Hans around this time. It is clear that she remained in London and did not attend either the ceremony or the reception, for in a rather curt letter her father reproved her, saying, 'In regard to Irene's wedding I know Mother has written you respecting it.' In a few brief words it was his way of showing his disapproval of her absence. He described the ceremony as 'a pretty little wedding', before digressing and congratulating her on getting a rise in salary at Crockers.

A few days later Will Johnson reproves his eldest daughter yet again, not only for failing to attend her sister's wedding, but also for being unwilling to meet Irene and Teddy in London the following weekend, when they returned from their honeymoon, telling her, 'I believe Irene thinks Hans was coming to visit you, and it was for *that* reason you were willing to cancel their visit. Whether this is right or not I don't know – this also may be a misunderstanding and should be cleared up.'

Amy may have been unwilling to face the embarrassment of attending Irene and Teddy's wedding. It is doubtful whether Hans would have been invited, or that he would have gone, even if he had been given a joint invitation with Amy. Secondly, if Amy had attended, with or without Hans, there would inevitably have been awkward questions on the lips of all the relatives. 'Is it you next, Amy?' or the unspoken question, 'After all, you and Hans have been together for over five years now, haven't you?' For Amy to appear before the whole family without even a token engagement ring

The newly wed couple - Teddy and Irene Pocock. (The Amy Johnson Estate)

would have been a tremendous humiliation, a blow to her pride that she would have found impossible to bear.

Irene and her father probably realised that Amy was now Hans's mistress, although it is doubtful whether her mother would ever allow herself to believe it. The reason why Amy had put Irene and Teddy off from visiting her in London during the Whitsun weekend was that she had already arranged to spend the holiday with Hans in the Hathersage area of Sheffield.

The rift between the two sisters, if one could term it that, could not have gone very deep, since Irene was soon involved in an endearing correspondence with her sister, prompting Amy to tell Hans, 'Irene wrote me last night - she seems to be perfectly happy and has everything she wants. She has a beautiful house, lots of nice new clothes, the car, as much tennis as she wants and lots of fun and good times. I am ashamed to confess that I find myself envying her, tho' I try not to, but I do feel lonely and frumpy.'

It was not only the Johnson family who were wondering when, or if, Amy would marry Hans, for a few weeks later Amy wrote to her lover telling him that she had heard from one of his sisters. Not wishing to miss an opportunity to push him off the fence he was so comfortably sitting on, she added, 'Fridy has sent me another little doyley for my birthday, like the one she sent me at Christmas . . . and she says she's going to send me one on each such occasion and *also to you*, until we have a set for our home. Don't be cross, I didn't say that, *she* did.' And as if to give an added push, 'I love the little doylies and am looking forward to having a set of them, aren't you darling, or don't you like your people to jump to such conclusions?'

Amy had been looking forward to Hans visiting her on her twenty-fourth birthday and had suggested that if he could manage a trip to London, they might visit the Hendon Air Display on the following Saturday, when the King and Queen would be attending. It seems that she had not forgotten the adventure of that first flight with Mollie eight months earlier. However, when the time arrived, business concerns took Hans away to Portugal and, not wanting to spend a weekend on her own, Amy invited herself over to the flat of her university friend Tuppy in Leighton Buzzard.

She felt that Hans now needed little excuse not to be with her, whereas earlier in their relationship he would not have missed such an opportunity to spend a whole weekend with her. Her schoolteacher friend was herself experiencing an unsatisfactory love affair, in her case with a married man, and no doubt a great deal of their time that weekend was taken up comparing notes.

As soon as Winifred Irving appeared on the scene in London to work in the welfare department of McVitie & Price's biscuit factory at Harlesden, she got in touch with Amy and the two women quickly renewed the friendship they had enjoyed during their university days. Such was their understanding and mutual admiration - they were kindred spirits - that Amy decided in the August of 1927 to move out of her West Hampstead digs to share a flat with her friend. They had much in common, both having three sisters and coming from families which were deeply committed to the Christian faith. Similarly, they had both rebelled against their religious upbringing and gone their own way.

Winnie had a single bed-sitting room in Bayswater on the Marylebone Road, quite near to Hyde Park, and was more than willing to share with her old friend since it helped out with expenses. To accommodate Amy, the landlady was prepared to rent them an extra room so that they could share a flat together. From here on, Amy's life appears to have become rather more settled, largely owing to her friend's calming influence. Winnie would read poetry to her of an evening, and generally spoil her when she was going through a bad patch with Hans. The two women had their own recreational interests: Winnie was a keen member of her firm's hockey team, whilst Amy was absorbed with tennis and regular visits to nearby public swimming baths. Amy also kept up her attendance at evening classes, which now included company law as well as cookery and German.

The two women entertained a constant stream of their female friends and siblings, who would do the theatre rounds when affluent or partner each other and dance to the latest records on the gramophone at home when less so. There would be Winnie's two sisters, either Belle from Paris or Elsie, one of twins, as well as Amy's sister, Mollie, and their mutual friends such as Gwyn or Tuppy from their university days.

There was a distinct cooling of the relationship between Hans and Amy towards the end of July. She replied to one of his less welcome letters thus:

> I must talk to you about a sentence in your letter which has already made a big difference in me. I do not know whether you intended me to attach to it the significance I am doing. The sentence I mean is, 'I don't want you to wait and see any longer.' I wish you would speak straight out to me, darling, instead of hinting at things in obscure phrases and expect me to understand what you daren't express. You're a coward in some things.

Exactly when and where Hans Arregger met Connie Richards remains a mystery, the most likely date being in 1927, or just before. Connie came from the Mossley Hill district of Liverpool and was described by a close friend as 'an attractive, blue-eyed blonde with a strong personality'. She was approximately six months older than Amy, and evidently a very bright young woman for she had graduated from Liverpool University in 1924 with a BSc. Honours degree in physics, and was subsequently awarded the prestigious Oliver Lodge Prize. She gained her Masters degree two years later, having already been appointed President of the Guild of Undergraduate Students. One can only surmise that Hans, who was keen on dancing, had met her at one of the university dances, since he and his business partner, Raymonde, had already established another office in Liverpool. At the time they met she was already working for the BBC as a programme organiser and occasional broadcaster, all traces of her Liverpool accent having been meticulously ironed out with elocution lessons given by her employer. Little did poor Amy realise what opposition she was now facing!

From the late summer of 1927 there were numerous hints in Hans's letters to Amy that suggest that he was playing a double game. There were references which Amy questioned with a great deal of probing, especially when he went so far as to mention that he went to dances at the local Palais, or when he told her of one of his 'Swiss clerks' whom he took out at weekends on his motorcycle. Amy's suspicions heightened further when in mid-September she returned to Hull for the first time since she had left home, so that she could be present at her parents' silver wedding anniversary, only to find that Hans was about to depart for a holiday in Scotland with 'some new friends'.

He seems to have kept up his correspondence with Amy during his time away, for she replied soon after: 'You will doubtless now have pleasant memories of Gleneagles. You must tell me more of your interesting conquest. Your explanation that she is not like *Gentlemen Prefer Blondes* . . . conveys absolutely nothing to me, as I'm not familiar with that book.' And then without even bothering to hide her vulnerability, she continues, 'Are you keeping up the acquaintanceship when you return to Hull, or was it just a passing interest?' Was he just teasing her

Amy became a keen tennis player after she met Hans. (The Amy Johnson Estate)

she wondered – he was fond of doing that, she knew. Amy just could not bring herself to believe that Hans could possibly throw her over after the length of time that they had been together.

From this point on their whole affair became extremely convoluted. He still wrote as regularly as ever with endearments such as, 'You've been a darling to write to me every day and to think of me so often', whilst she attempted to arouse his jealousy by telling him of various 'charming boys' that Winnie was introducing to her, some real and some imagined.

In October, he returned from a business trip abroad and whilst passing through London called to see her, bringing chocolates and cigarettes. One gathers that he was still hedging his bets and was not prepared to spurn the love and adulation she lavished upon him, for soon after his return to Hull, she wrote:

> You've been as nice as you could possibly be to me the few hours you've been here, darling. Winnie said she liked you a lot more this time and I asked her why and she said it was because you were so pleased to see me. I did love to see you and want to see you again soon. Maybe I shall. I love you darling, such lots and lots. All my kisses.
> *Ta chérie.*

This is hardly the language of a woman who is expecting to be cast aside!

Vernon Wood's secretary, Miss Roberts, left Crockers in September in order to be married, something which women were expected to do in those days. This left Amy to take over her position with a salary of £4.10.0 per week. Just before she started in her new post, Will Johnson took the opportunity to give his daughter some fatherly advice.

> I hope you are getting on well at the office, dear, and that you still like the work and feel equal to it. Very glad indeed to know that you are Mr Wood's secretary and that you've got a rise in salary, which I've no doubt you can do with. My word Amy, you are jolly lucky to jump into such a position . . . I therefore hope you will do your part in every respect to hold it and to give every satisfaction. There is one thing I know about Mr Wood, and that is he is a terror for punctuality, so I hope you will keep your eye on this so that he won't have any cause for complaint respecting it. He has a system of knowing . . . Some girls may think they can just steal a few minutes in the morning and again at lunch, but I know that Mr Wood gets to know all these points.

Amy must have heeded the advice, or known a way round it, for within a few months she received another ten shillings rise in wages, and was able to pay back some of the money she owed to Hans.

Throughout the rest of the year, realising that serious opposition might well be in the offing, she attempted to revive her flagging relationship with Hans by referring to a string of flirtations that she was supposedly enjoying. She told him that she had discovered that she had a 'secret admirer' in one of the flats where she was living, and when this drew little response, she made a lightning and unannounced visit to Hull, no doubt hoping to see for herself exactly how matters stood between them.

Her visit gave her little cause for alarm, and if anything reassured her that their relationship was back on course, slightly wavering maybe, but still holding. This impression was strengthened in Amy's mind when, upon returning to London she found a letter from Hans, written on the evening before she made her trip. She wrote back:

> When you wrote those nice words at twelve o'clock on Friday night, you never guessed I was still up getting ready to come and see you, did you, darling! I did enjoy surprising you as much as anything and you really did seem glad to see me. Write to me soon, darling. You were a dear to me this weekend and gave me a nice time. Am looking forward to seeing you again. All my love, thoughts and kisses.

In the light of Amy's visit and subsequent events, one cannot help but come to the conclusion that Hans's vacillation amounted to a form of mental torture, whether intentional or not. He could not have made his position clear to Amy when she visited Hull, and even in January 1928, when he came one small step nearer to transparency, he clouded the issue, causing Amy to write, 'I wonder why you do so enjoy teasing me, especially now when I don't get so insanely jealous. I am of course curious as to your new acquaintance – which will probably be a dog or something of the sort, I expect – but I'm afraid I'm not a wee bit concerned about it.' And then in an attempt to salvage her pride, whilst at the same time being unable to curb her curiosity, she continues, 'You forget I've changed and am much more sensible now. All the same it will be interesting to know more on the subject as I don't at all understand the abrupt way in which you introduced your "new acquaintance". You didn't even warn me there was one.'

Just before Christmas, when Amy was due to arrive in Hull to be with her family, she was annoyed to find that Hans was spending most of his holiday with friends at Cottingham – or so he told Amy. She accused him of living two lives, 'one in which I share and one where I might as well not exist'. However, upon his return to Hull he did manage to see her again during the holiday period and made an attempt to mollify her before she left for London. He accompanied her to the city's Paragon station to see her onto the train, but not before giving her an envelope which she later discovered contained a generous gift of money. Whether this was a means of easing a troubled conscience or not one can only conjecture; but it was a rather strange and cold way of giving her a Christmas present.

It was an opportune time for Amy to have some spare cash, for when she arrived back at the flat, she found that Winnie was in debt. It appeared that she had not been paying their rent, which meant that Amy had no alternative but to finance her for the time being. Not wishing to be separated, they decided that in order to economise they would move into a hostel near Oxford Circus. Their plan was to save money by renting two wooden-partitioned cubicles there at 9 shillings each per week, thus allowing Winnie to pay off her debts. Amy also decided that it would give her an opportunity to save enough to purchase furniture and a flat of her own. However, their plan backfired and they stuck it for only five weeks. They found that at weekends they had to either sit in a cheerless, cold lounge or lie in their beds in order to keep warm; the only alternative was to go out for entertainment, which meant spending more money.

As restless as ever and determined to stick together for mutual support, they moved yet again, this time to a rather miserable-looking house in Maida Vale. Its sole redeeming feature was that their small rented room overlooked the pleasure gardens and tennis courts opposite.

Although there is no doubt that Amy had come to love London and its theatres, concert halls and parks, and would have settled there if marriage had been forthcoming, the constant urge for travel and adventure never left her. A quote from one of her earlier letters to Hans, written when she was in one of her 'dog days', illustrates this quite clearly, as well as having a rather prophetic ring to it. She wrote, 'I just hate streets, houses, tubes, buses, people, stuffy rooms and typewriters. I wish I could climb a mountain and be high above all these things – where there's light and air and sunshine.'

This must have been exactly the kind of mood she was in when, on the spur of the moment, she took leave from her job one Friday evening in March 1928 and hurried to Paddington station to board a train bound for Cornwall. A young lad who sold eggs to some of the staff at Crockers' office had told Amy about one of his relatives who ran a small private hotel in Newquay, and of the delights of that particular stretch of coastline. Almost immediately Amy's mind was made up – things would never be quite the same again.

No sooner had she arrived in Newquay than she was posting a hurried note to Hans telling him of the wonderful view from her bedroom window, almost identical to the picture on the holiday postcard she was now sending him, and adding, 'But it doesn't show the intense blue sky, green sea, white seagulls and the brilliant sunshine.' And then as if to add to her bliss, she instinctively asked, 'Can't you come next weekend?' It is obvious that her feelings for him had not changed and she still held out the hope that they might recapture the love that they had once shared.

On the last evening of her stay in Newquay, she told Hans of the long walks she had enjoyed and of finding her way around without getting lost, in spite of his contempt for her 'bump of direction' – or lack of one. 'I've gained quite a reputation in this hotel for my ability to find my way – which isn't easy in Cornwall.' And then, describing a walk from Perranporth:

> I walked along the sands till I found a sheltered spot for my lunch. I'd really intended to have a quiet day because I'd walked about fifteen miles the day before to Bedruthan Steps and nearly all the way back. However, after my lunch and a quiet read, I felt so little tired that I decided to try to find St Pirran's lost church in the sand dunes. I walked a long way along the sands – firm and hard and yellow and utterly deserted except for seagulls . . . then I turned inland and found the church with little difficulty. The sand dunes were inhabited by hundreds of rabbits and it was almost possible to imagine oneself in the desert when down in the hollows of the dunes. The sky was intensely blue but it was bitterly cold, although all the time I have been out walking without hat or coat – just carrying my mac in case of need.

Her sense of adventure and ability to chart her way across unfamiliar terrain was soon to be tested in a much more dramatic way.

Away from the hustle and bustle of London and all its distractions, Amy's brief holiday in Cornwall had enabled her to clarify her thinking and unscramble the confusion of her mind with respect to her relationship with Hans. During her solitary walks along the stretches of Perran Beach and across the inland sand dunes, she had had time to reflect and ponder the course her life was taking. It became clear to her that the state of indecision she was in could no longer be allowed to continue. It was almost two weeks after her return before she even attempted to write to Hans again, and her incisive and lengthy letter, which was eventually to untie the Gordian knot, was the result of much deep thinking, and was no doubt drafted and redrafted several times before she sent it.

> Dear Hans
> Now I must try and explain as clearly as I can the lines on which I have been thinking for some time and the decision I have come to. This is rather difficult and I hope you will hold in check any irritation you may have for things I say, until you reach the end of the letter. My life up till now has occupied two volumes – (1) my childhood and school days, and (2) you. The first one finished when I went to Sheffield and it is now time to close the second one. From the age of eighteen until now, when I am nearly twenty-five – practically seven years you've been the

predominating factor in my life. I don't blame you for that – it was I who was the fool – and now that I'm so much older and more experienced I can look back on those years and see how utterly stupid I have been. I think I am now strong enough to cut it all out and that's what I intend to do. No more looking for letters that came less and less often; no more wondering when you'll come to London again to see me, or rather when you're on business; no more puzzling over the question of how much you really love me and what are the thoughts about me that must sometime be going on in your mind. I do not think there is any need for us to continue writing to each other – it is obviously becoming more and more of an effort on your part and the attempts you make to fill up a sheet of notepaper are not worth the time and trouble involved.

Let's go back to the beginning of all the trouble and look at things plainly. You made the first mistake by treating me in the way about which you once said, 'If a chap had treated one of my sisters like this I'd have killed him' – ironical isn't it? I've had no brothers to help give me a saner, healthier idea of the relationship between the two sexes, or to protect me in the good old-fashioned way. I wonder what made you say such words out loud – they've excused you for a lot in my eyes, even though you resolutely forgot them as soon as uttered.

I made the next mistake in allowing you to treat me in such a way. But whereas you had no excuse at all for your actions, I can consider I had a good deal. First, you were practically a god in my eyes and everything that you did was right just because you did it. Secondly, I knew practically nothing about men and the way they differed from women and I was ashamed of my ignorance and tried to cover it up by pretending to know everything. The few ideas I had were totally wrong and absurdly ridiculous. I can't help but smile now when I remember some of them. In a vague sort of way I felt that in yielding to you I was losing your respect for me, and yet my mind was full of unsolved problems, wonders and perplexities. I awfully wanted to find out things and have actual experiences, and being *you* who was there and ready and wanting to give me them, just clinched the matter. The more I got to know the more I wanted to know. See where I am now – the matter no longer interests me. I feel I know all there is to know – that's all men want with girls, you can see it all around you – in the painted faces of nearly all the girls you meet and in the staring desiring eyes of the men who pass you in the street. How I loathe them all and how I despise myself for being just the same as they. But please don't mistake me – I owe all my knowledge to you and I am grateful to you for your training and for the worldly knowledge you have given me. If I could, I wouldn't wish again to be the stupid little innocent I was when you first knew me. I used to think it was *hurting* you and I was so sorry for you and felt so tender towards you because I thought you were hurting yourself for the sake of giving me enjoyment. Oh, isn't it too funny.

Ever and ever so many times I wanted to try and undo the harm, but I wasn't strong enough and I used to think about it for hours and try to justify my conduct. First I thought just to love you so much was sufficient excuse, and then I thought out all the ethics of marriage and sided with the view that it was merely legal licentiousness. I formed my own little code of right and wrong - that, because we live in a wide tolerant age, the mere form of marriage makes no difference to whether it is right or wrong to be a man's mistress. I decided that there was no wrong being done at all if the lover and the mistress really and truly cared for each other sufficiently to be satisfied with each other, and never to go to anyone else for their pleasures, and that the form of marriage made no difference at all except to give legal recognition to the children, if there were any.

This, I remember, lasted me for a long time and then I began to find holes in it. It was one day after you'd really said something fairly definite for you that you thought some day we might actually get married. I then dropped all my former line of reasoning and merely based justification for my conduct on the grounds that we should really be husband and wife some day and there was therefore no sense in waiting until the day and we might as well carry on as we were doing.

The last phase and the one I hold at present, is that there is no justification at all, except that everyone is more or less weak, and that the less weak should help the more weak. This is of course seldom carried out and everyone is despisedly weak when it really comes to those moments of intense physical feeling which, I suppose, the good benevolent Lord endowed us with for good ends. Pity he didn't see the bad ends as well before he pronounced his work good.

You will perhaps be wondering what all this has to do with the present state of affairs and my new decision - just this. All the time that you have known me you have found so many things to grumble at and find fault with. As I see it now, the only thing which has kept us together for so long has been your intense desire, and my reciprocating desire, for sexual intimacy (loathsome words). Apart from this one thing, the whole of your instincts seem to be against a married life. This has been all right for you, so long as I was content to let things drift on and on as they were. I've been for the greater part of seven years content to let them so drift because I thought that there was something at the end of it. Always I've had before my eyes the picture of a home where there'd be no outside interference causing friction, and where I'd be surrounded by things which would all belong to us, for which we'd worked hard and paid for and chosen, and where there would be children of my very own whom I'd train up so carefully. This is what I've always wanted so much and you, being the only man in my life, were naturally the one to whom I should look to give me these things - apart altogether from the urgent desire I've always had for you yourself.

Now, however, I can no longer persuade myself into believing this is a possibility - it is too obvious that in every way you are getting to care less for me. Everything shows it. You don't even want the sexual love now. Your letters come less and less often . . . shorter . . . less affectionate. They are now merely scrappy news bulletins prefaced with 'Darling' – the sugar coating the pill. How simply farcical it sounds now when so obviously you don't mean it. I have told myself once or twice, that it was most unlike you to write it if you didn't mean it, until I realised that you couldn't address me as anything else because to do so would in some manner bring you face to face with your change of feelings. I've noticed several times your tendency to let things drift as long as possible, sooner than stop and face them.

It is always I who stops the drift and often I have done so foolishly. I can see it now. I wish to God I'd never mentioned the subject of marriage to you. I was a fool and have, I can see, defeated my own ends, but it seems that it was inevitable. I was impatient to know what was going to happen to my life and you irritated me so tremendously. You have often reminded me of the donkey I once took Betty for a ride on at Bridlington. It went slowly and I was eager to get back to my games and the more I urged and coaxed the slower it went, until finally it stopped, planted its legs firmly on the ground and refused to budge. I pushed it and pulled it and almost got to kicking it I'm afraid, but no use. Finally, I lifted Betty off and we turned round and walked back. The donkey followed a minute later.

There is a difference however in this letter from the others where I've written to you on what was to me the most important thing in my life. I'd better point it out in case you don't realise it for yourself. Always before, I've made many and varied protestations of my deep love for you and told you frankly how much I wanted you (not sexually, as you probably thought) and how sure I was we could be happy together, and begging you to be quick to come to London. Now I no longer love you, I no longer want you, sexually or any other way. I don't believe we could for a single moment be happy together, and if you come to live in London I should probably leave – not because I couldn't bear to be in the same city with you – I don't hate you, I'm just totally indifferent . . . I'd always be afraid of meeting you and that would get on my nerves.

I feel I have now written enough to give you some indication of my thoughts, but I just want to say two more things.

One, that I don't want this to affect my relationship with your people. I love them all very much and even though they probably only care for me for your sake, I still feel I want their friendship so much. The second is that I still owe you £3.0.0 which I cannot send you at the moment, I'm afraid. I'll send it along as soon as I can. I am very grateful for the loan and for the way you've always helped me in money difficulties.

As a last thing I want to tell you something which I can't put plainly (there would be no need to tell you at all if you could read between the lines, but you're not good at that as I've found before). I'll tell you in allegory.

In a certain room of a certain house a beautiful fire had been kindled by a youth who lived there. It was his duty and his pleasure to keep it always burning brightly, and this he loved to do. But gradually he got somewhat tired of always picking out the best pieces of coal and wood and he grew careless and threw on anything handy until by and by the fire grew duller and got choked up by ashes. Then there came a day when it was practically out and only then did the boy grow alarmed. As he arose considering what to do, a voice whispered, 'Just leave it, that's the easiest way. It will go out and if you want you can light another. But if you want this one to brighten up again you must do something at once and look after it continually. Another time it will go out more quickly.'

It is stupid I know to make up such a silly story. Oh well, once a fool always a fool I suppose.

I hope you will soon succeed in getting a good job [he had decided to give up his business] and I wish you every success and lots of good luck. Amy.

And then more in blind hope than the cold reality of the situation, she added a postscript: 'It isn't of course possible that I have totally misjudged you?'

She need not have bothered to waste the ink to add those final words, for in spite of another meeting with Hans at Easter, during which they came to 'a new understanding' that they would from now on just remain as friends, she had assessed the situation with perfect clarity. Another fire had indeed been lit by Hans, one which was to thrust Amy into a new-found interest.

The serenity enjoyed by those living in Castellain Road was such that the peace of a summer's weekend was disturbed by nothing more than the gentle and soothing sound of ball upon racket coming from the tennis courts opposite. However, this tranquillity lasted only until weekend flying caught on in the late 1920s. Then the residents soon discovered that they lay directly under the flight path of the ubiquitous de Havilland Moths which flew above the straight line of the Edgware Road nearby. Both pilots and their pupils used it to navigate to and from their small grass airfield at the London Aeroplane Club, some 5 miles away.

The exhaust noise from their Cirrus and Gipsy engines was an irritation to most local people. There were exceptions however, one such being the young white-clad tennis player, who paused before serving to cup her hand over her eyes and squint into the sky. To her the purr of those engines was a siren call to the adventure for which she longed. It was one she could not resist.

Amy's almost forgotten interest in flying had been revived in the late April of that year with the release of the much-publicised film *Wings*, an epic which was currently being shown in London. The film set new standards in aerial photography by giving dramatic cinematic effect as it portrayed air battles over France in the First World War. To heighten the excitement, it contained the added ingredients of love and romance provided by Hollywood stars, Clara Bow and Gary Cooper. Coincidentally, the annual Hendon Air Display, due to take place on the last Saturday in June, was being heavily advertised and after seeing the film Amy was spurred on to purchase a ticket in advance.*

In spite of forgoing her visit to Hendon in favour of the tennis tournament, Amy wrote off to the de Havilland Aircraft Company for their prospectus, but their fee of 5 guineas an hour for flying lessons was quite beyond her means – more in fact than she earned in a week. Even with this disappointment, her recent craze for flying and aeroplanes did not waver.

On the last Saturday afternoon in April, when Winnie was out playing hockey and had left her flat-mate on her own, Amy decided on the spur of the moment to board a bus going in the direction of the Hendon aerodrome, in the hope of seeing some of the magical machines at close quarters. From the open top deck of the bus she could see the small biplanes coming lower and lower, their leather-helmeted pilots and pupils clearly visible. It seemed to her, as she made for the steps to alight from the bus, that their wheels would almost brush the top of her head as they glided in to land.

Not put off by a large sign reading, 'London Aeroplane Club. Private', Amy sauntered into the entrance of the aerodrome and walked towards the small wooden clubhouse. Several club members were sitting in front of its verandah in deck-chairs, engrossed in watching the machines landing and taking off. Stag Lane had four small adjoining hangars bordering an undulating grassed area, which was hemmed in on all sides by rows of suburban houses. Apart from the proximity of local housing, it was hardly the ideal location for an airfield, since it was low-lying and therefore prone to autumn mists and winter fogs. Amy took a vacant seat alongside some of the members, who seemed far too busy chatting to notice her, and hoped that her unauthorised intrusion would not be challenged.

After watching for a while, she plucked up courage and asked one of the pilots standing nearby how much it would cost her to learn to fly. 'Two pounds an hour for instruction. Thirty shillings an hour solo. Three guineas entrance fee and three guineas subscription.' And then as an afterthought, he added, 'Takes from eight to twelve hours to learn.' Amy realised that the possibility of tuition was now within her reach and promptly asked, 'Could I start straight away?' No, he replied, she would have to apply to be elected,

---

* As it later transpired the air display clashed with the local tennis tournament in which Amy had qualified for the women's doubles and she forfeited the ticket.

which meant waiting her turn to become an associate member of the club.

In her memoirs, Amy described this day as 'the most important milestone in my life'. From that point on she pursued an ambition to learn to fly with a greater tenacity than that shown in the pursuit of her Swiss lover. This time there was to be a more successful outcome.

Owing to the large numbers applying for flying tuition at that time, there was a delay of several months before Amy's application was finally accepted. During this period Amy impatiently wrote several letters to the club's secretary, urging him to review her application. This did little to help her cause and was not appreciated by the rather pompous and curmudgeonly Harold Perrin. He was not only the secretary of the LAC, but also of the Royal Aero Club itself, an organisation which controlled the five state-aided flying clubs based in London, Durham, Lancashire, the Midlands and Scotland. The Government was anxious to promote aviation in the UK, hence the reasonable rates for tuition, which surprised Amy and made it possible for her and many others to learn to fly.

Whilst Amy was waiting to hear when she could commence her flying lessons, Hans visited her quite unannounced and they went out together for dinner. (It was the first time she had seen him since the Easter.) Amy returned alone much earlier in the evening than Winnie had expected, looking extremely shocked and shaken. Hans had married his 'broadcasting friend' on 14 July in Hull. Winnie remembers Amy throwing herself onto her bed sobbing inconsolably. The butterfly was about to emerge from its chrysalis!

Chapter Seven

METAMORPHOSIS

There is no doubt that when Hans broke the news of his recent marriage to Connie Richards that Amy was more than devastated, for as she was later to admit, she was suicidal. In spite of all the prevarication and stalling that Hans had displayed for the last twelve months, she could never really bring herself to accept that their relationship was finally over. She had always believed that she could lure him back somehow. If not by provoking his jealousy with stories of all the young men she was now dating, then maybe by taking up what was considered by most people at that time to be the dangerous sport of flying. Even Hans's sister Fridy conceded some forty years later that: 'We were of course sure when he brought Amy home that he would marry her.' The blue vase, Swiss pictures and gifts that she had been given as wedding presents by his Swiss friends and family, and which she had been loathe to dispense with, now remained to mock her.

It was just as well that Winifred Irving managed to persuade Amy to keep to her promise of accompanying her and Gwyneth Roulston on their planned walking holiday commencing in the French Alps that August in 1928. The fresh mountain air, healthy exercise and companionship she enjoyed during that fortnight helped her to throw off some of the depression she was experiencing. The trio, clad in appropriate walking gear, spent four days' hard trekking from Grenoble to arrive in Turin, and from where Winifred took the train back to Paris to stay with her sister, Belle; whilst Amy and Gwyn travelled on to Lucerne, where Amy took the opportunity to visit Hans's family. It says much for her integrity and strength

Postcard sent by Gwyneth Roulston on 13 August 1928 whilst she, Amy and Winifred were on a walking tour through Italy and Switzerland. Amy took her first flying lesson five weeks later at the London Aeroplane Club, Stag Lane, Edgware. (Patricia Stewart)

of character that she maintained such a good relationship with the Arregger family, in spite of the disappointment she had suffered.

Upon returning home it was not long before a letter arrived from the London Aeroplane Club to inform her that she could commence her flying lessons. From now on her life would never be the same again.

The 30-foot wingspan DH Cirrus II Moth biplane in which Amy first learned to fly stood resplendent outside the Stag Lane hangar, its silver-doped wings and tailplane contrasting with its canary yellow fuselage. The Cirrus Moth was the forerunner of the classic British light aeroplane of the 1930s, the ever-popular Gipsy Moth, and like many of its company's successors, it derived its name from the fact that its designer, Geoffrey de Havilland, was a keen lepidopterist.

The Cirrus version of the Moth had been designed with the private owner very much in mind, and to this end it incorporated twin mainplanes which could be folded back neatly to the fuselage so that it could easily be garaged when not in use. Although it was of wooden construction it was extremely sturdy, and was powered by a reliable 80 hp engine as simple as that found in the average motor car. The controls in the two open cockpits were dual-linked for instruction, and communication between pilot and pupil was by means of a speaking tube connected to headphones. The instrumentation layout of the cockpit might well have appeared baffling to Amy at first, but it was in fact quite rudimentary, consisting of a centrally mounted compass; air-speed, rev-counter and turn-and-bank indicators; oil-pressure and temperature gauges; and throttle. Their function and use was carefully explained even before she climbed into the aircraft.

Amy's first flying lesson lasted for only thirty minutes and took place on Saturday, 15 September, under the tutelage of the youthful junior instructor, Captain F. R. Matthews. Unfortunately for Amy, she was given a flying helmet several sizes too big for her, which meant that its integral headphones came down well below her ears, with the consequence that she only half heard the various instructions she was being given.

The man who told Amy she would never make a pilot – Captain F.R. Matthews, her first Flying Instructor. (Ron Neudegg)

The aim of the first lesson was to familiarise a pupil with use of the aircraft's two major controls, the joystick and the rudder bar. This meant the pupil holding the stick very lightly and gently following the dual-controlled movements made by the instructor, whilst at the same time observing their effect upon the attitude of roll and pitch of the machine. Similarly, this procedure would be repeated with the pupil's feet placed lightly upon the pedals, so that the effects of rudder movements on the yaw of the aircraft could be seen.

Owing to the ill-fitting helmet, Amy did not obey the instructions she was given, and Matthews, who was not the most patient of men, was no doubt fulminating in the rear cockpit, thinking he had a no-hoper on his hands. After landing, he told her that she would be wasting her money if she continued with instruction, and that she was unlikely to make a pilot. Little did he realise the stubborness and determination of this small woman. She was intelligent enough to realise the cause of her poor performance, and booked her next lesson undeterred.

Amy only had a further five lessons during 1928, some with Matthews and some with the more mature and sympathetic Chief Flying Instructor, Captain Henry Valentine Baker, MC, AFC, before she even mentioned to her parents that she was now taking flying lessons. Much to her surprise their response was rather muted, and one presumes that they thought it was just another five-minute craze that their daughter was going through.

Amy's next lesson did not take place until January 1929 and one can only conjecture that the reasons for the delay and discontinuity in instruction – not a good thing in itself – were two-fold: a shortage of money and the unsuitability of the weather for flying in the early winter months. The

Captain Henry Valentine Baker, MC, AFC, was one of Amy's instructors at Stag Lane and at Heston Air Park, 1928-1930. (Martin–Baker Ltd)

Captain Herbert Gardner Travers, DSC., (facing the camera) instructing pupils,
Bill Oliver and Phil Cooper at Stag Lane, circa 1931.
(Universal Pictorial Press & Agency Ltd)

London Aeroplane Club did have its social side, however, which centred largely on the club's bar, where flying stories would be the main topic of conversation, and Amy participated whenever she could. In the depth of winter there would also be the occasional diversion such as ice-skating on the Elstree reservoir or some other outing in London.

Eva Fitzpatrick, whose father, Major Herbert Travers, became Chief Flying Instructor at Stag Lane, gives us a first-hand glimpse of club life at that time in her nostalgic book, *Cross Country*:

> Although the majority of pupils at the London Aeroplane Club were men, there were a scattering of girls there. Some of the girls were not, in fact, pupils, but were 'me-too's' – such a withering collective noun for the girls who could not afford flying lessons or who, perhaps, simply wanted to see what was going on and be a part of the general flying movement.

Eva also remembered some of the women pilots who were there at the same time as Amy, or just after, including Pauline Gower, Dorothy Spicer, Winifred Spooner, Joy Muntz and the celebrated Jean Batten, all members of a select band known at the club as 'the Beauty Chorus'. Although members were predominately from the wealthier sections of society and of a higher social standing than Amy, including Lady Mary Heath and Lady Bailey, as well as celebrities such as the film actor, Will Hay, snobbery never seemed to be much of a problem amongst the flying fraternity.

Once the flying bug had well and truly bitten, Amy found it hard to maintain the interest she had once had in her work at Crockers. The long hours now became even more irksome and she was overcome by boredom, since she saw her employment as nothing more than a means to an end, merely to provide enough money for flying lessons. She longed for the freedom of the open skies and lived for the weekends, when she could visit Stag Lane and be amongst her new friends. Dancing, the cinema and the theatre were now firmly relegated to the background. She needed all the money she could muster, especially now that she had decided to purchase her own car, a two-year-old maroon Morris Cowley, with a £70 loan from her father.

At this stage one must ask why Amy was not content to achieve the distinction of a private pilot's licence, and then to enjoy the social side of her new interest. What gave her the drive to do more than this, to achieve the fame she did?

Her motivation derived from several sources, including the two obvious ones: her disappointment in love – she would show the man who had treated her so shabbily what she could really do; and her natural desire for adventure. Undoubtedly, her role models would have been the two women who had already achieved distinction as solo fliers: Lady Mary Heath, who had the honour of being the first woman to make a solo flight from the Cape to the UK; and her Irish contemporary, completely opposite in character and temperament, Lady Mary Bailey, the first woman to fly solo *to and from* the Cape.

Amy would have had no need to read about these two remarkable women in the newspapers, for their historic flights had occurred just before she joined the LAC. One can be sure that their exploits would have been a regular topic of conversation in the clubhouse at Stag Lane.

Lady Mary Heath had been the very first person to receive flying instruction at Stag Lane when the subsidised club opened in 1925, at that time she was Mrs Sophie Mary Elliott-Lynn, a rich young widow from Limerick. She was probably the most colourful aviatrix of her time. Her wealth, glamour and imperious manner, allied to an aggressive and gutsy feminism, meant that she was not a woman to be overlooked, especially by the press. One of her many adventures involved a successful parachute descent. Her first attempt had been frustrated when, just as she was standing on the wing and ready to jump, the aircraft developed engine trouble. This left her in the precarious position of clinging on to the side of the plane as it made a forced landing.

Unperturbed by this brush with death, she set about showing that women could compete on an equal footing with men in the world of aviation – something that would have appealed particularly to Amy. Setting off before dawn, she flew her Avro Avian biplane a distance of 1300 miles in one day, visiting every known aerodrome and landing ground in England south of Manchester, before flying on to Newcastle, thus making seventy-nine landings before nightfall!

She married the elderly Sir James Heath in October 1927, and they set sail for South Africa with her Avro Avian stowed on board. It proved to be a loveless marriage, in which she was expected to be little more than her husband's nursemaid. Before long she was bored to distraction, and as a means of escape she managed to cajole her husband into agreeing to her flying back to England. She arrived at Croydon in May 1928 after a solo journey lasting three months, showing a shapely leg as she stepped down gracefully from the open cockpit of her aeroplane. To the astonishment of the press she was dressed in furs and a close-fitting, black straw hat, and looked as if she were about to take afternoon tea at the Ritz. No doubt the value of the publicity she courted and received was not lost on Amy.

Lady Mary Bailey, the daughter of an Irish peer and wife of the wealthy South African magnate, Sir Abe Bailey, took a less frenetic attitude to flying. Although both women were often referred to as 'the flying aristocrats', she was, in fact, a modest and unassuming middle-aged mother of five children. She too had learned to fly at the LAC and acquired her 'A' Licence (equivalent to today's Private Pilot's Licence) a year after Lady Heath. Being a competent pilot who owned a 'go anywhere' DH Cirrus Moth, she thought it no big deal to fly down to Cape Town to be with her husband. She took off from Croydon early in March 1928, without any intention of setting or breaking any records, and arrived after fifty-two days, greeting her husband, 'Hello, Abe, how are you? I'm a bit late, but I got muddled up in the mountains.'

After spending almost five months in South Africa, she decided to make the return journey to the UK, and this time took a leisurely four months to complete her flight, a round trip of 11,000 miles which was described by the Secretary of State for Air, Lord Thomson, as 'just pottering around Africa'. Pottering or not, she was awarded the Britannia Trophy in 1929 for what one correspondent considered to be the greatest solo effort of its time.

Each one of these pioneering flights left an indelible impression upon the young novice pilot, Amy Johnson.

The weather conditions improved sufficiently in January 1929 for Amy to resume her flying lessons with Captain Baker. She had mastered the elementary art of keeping the aircraft on a straight and level course, as well as making gentle turns, but now came the trickiest part of flying, making a smooth three-point landing, in which the two main undercarriage wheels and the tail-skid touched the ground simultaneously. It was a manoeuvre which required a fair degree of skill in making a correct approach to the airfield. One needed the right amount of height, and a sensitive co-ordination of hand and eye in judging the exact point at which to virtually stall the machine onto the grass without it bouncing. With most aircraft of that period, as soon as the nose of the machine lifted just before the touch-down, direct forward visibility disappeared, particularly in the Cirrus Moth, since the four cylinder heads of the engine were in line with the top of the fuselage.

Placing an aircraft gently onto the grass never seemed to come naturally to Amy. Her flying log is littered with comments such as, 'practise landings', or 'bad landings', in the early part of the year, and even after she had gone solo. Not mastering this vital part of her tuition must have really worried Amy, as is evident in a letter to her mother early in March from Brentwood Place, Golders Green (she had moved from Maida Vale to be nearer the airfield), when she wrote, 'I'm persevering with the hypophosphates, but can't say they are doing me much good.'*

Captain Baker left Stag Lane in April to take up a similar and presumably a better-paid post as the Chief Flying Instructor with the nearby Heston Aero Club. It was then in its infancy, but went on to enjoy a reputation for catering for the *bon chic, bon genre* of the flying world, with a customs post, its own small hotel and restaurant, as well as floodlighting for night flying.

Baker's replacement was the tall and amiable Major Herbert Gardner Travers, DSC, a fair-haired, slightly balding man in his late thirties. He had

---

* A fairly common remedy during the 1920s.

learned to fly at Hendon in 1914 before serving with the Royal Naval Air Service during the First World War, and had subsequently been employed as a test-pilot with the Blackburn Aircraft Co. at Brough and overseas, before becoming a flying club instructor.

Herbert Travers was the perfect man for the job at LAC, for he had that vital ingredient in an instructor, a great concern for the safety of his pupils, so much so that to the less discerning, he might have appeared to be over-cautious. This attitude is well illustrated by one of his favourite maxims, 'Always have the courage to turn back.' Dressed in his customary plus-fours and tweed jacket, he would arrive at the airfield in Jemima, his much-loved Armstrong-Siddeley open tourer, and whenever he thought flying conditions were unfavourable, which usually meant that if the church spire on Harrow Hill was not clearly visible, his comment would be, 'Today no see, no fly.' The disappointed pupils would then amble back into the clubhouse to commiserate with each other and to curse the vagaries of the British climate.

Amy had clocked up almost twelve hours of instruction before Travers began to coach her to improve the quality of her landings. He knew from discussion with Matthews that she lacked skill in this area. They did not get on, but he must have inspired some degree of confidence in her, for on 9 June she made her first solo flight, after a total of fifteen hours and forty-five minutes of instruction. Ironically, it was Matthews who climbed out of the cockpit on that day with the casual remark, 'OK, Miss Johnson, away you go. Make one circuit of the airfield and then come in and land.'

Amy was in the air for what seemed an interminable five minutes. The butterfly had fluttered its wings! Within a few weeks she was awarded her 'A' Licence, having passed all the necessary flying and medical tests.

With the added confidence of gaining her pilot's licence, Amy was now fired with a new enthusiasm: to become a commercial pilot and have a career in aviation. She became determined not only to achieve her 'B' Licence so that she could carry passengers, but also to qualify for her 'tickets' as a ground engineer in both aero engines and airframes. She began to wander into the hangars where the aircraft were serviced, an intrusion which was frowned upon. However, such was her zeal and thirst for knowledge of what went on under the engine cowling of a Cirrus Moth that she soon struck up a friendship with the men working on the club's machines.

The Chief Engineer at Stag Lane was the placid and easy-going Charles G. H. S. Humphreys, known to all the club members as Jack. He was a small and energetic man with a neat Charlie Chaplin moustache. He took an immediate liking to Amy, and was mildly amused when she began to ply him with questions on magnetos and carburettors, surprised that she had such knowledge, something she had acquired since owning her Morris Cowley. Their friendship developed until, by mutual consent, she was arriving at the hangar an hour or so before she was due at her office. She would then don a pair of grimy overalls and do odd jobs to help the two fitters who worked alongside their boss.

It was in this unladylike environment that Amy gained valuable 'hands-on' experience to reinforce her studies for her ground engineer's licences, and where she first earned the nickname 'Johnnie', one she enjoyed. It was not long before the staff at Crockers were noticing that her normally well-manicured hands were no longer quite what one expected of a solicitor's secretary.

The latter part of July 1929 had been unusually hot and oppressive, with widespread thunderstorms throughout most of the British Isles. On the last Sunday of the month, Amy received an urgent call from Hull to return immediately; there was serious trouble at home.

Teddy Pocock had on gone on one of his firm's Saturday outings alone, Irene having chosen to stay at home. Her younger sister Mollie had visited her early in the afternoon, when everything appeared to be normal. However, upon Teddy's return to the main railway station around midnight, a member of the Johnson family met him to warn him of what had happened. Irene's maid had visited the house late in the evening, to find that Irene had sealed all the windows and doors of her small terraced home and then knelt down beside the open door of the gas oven with the tap switched fully on. She was found unconscious and, although a doctor was called immediately, she failed to respond to resuscitation. She left a farewell note to her husband requesting, amongst other things, that no mourning black should be worn for her.

One is left with nothing more than speculation as to why a young woman who had only recently been married should take her own life. At the inquest, Will Johnson, who at the time of her death was away on an overseas visit, could only recall one occasion when his daughter had threatened to take her life, and that was when she was studying to be a schoolteacher, some six years earlier. She and Teddy appeared to have been happily married, apart from the usual squabbles and tiffs between young newly-weds. They had no financial worries. Both were in reasonably well-paid jobs, and they made extra money from their involvement in local theatricals. There is the possibility that Irene could not face the thought of becoming a mother as well as a wife, as she enjoyed a fairly hectic social life. The only other faint clue suggests that there may have been a rupture in a close family relationship somewhere, one which might have disturbed Irene. This comes across in a letter from Amy to her mother, written a month before Irene died, in which she says that she has finally decided to leave Crockers and take up a full-time career in aviation. She then goes on to add somewhat cryptically, 'Sorry to hear about Irene. I'm sure there must be a misunderstanding somewhere.'

Irene's tragic death had a devastating effect upon all the members of the family, both parents and children alike. The fact that Will and Ciss Johnson's eldest daughter was now determined to embark upon a career in aviation, with all its attendant dangers, only made matters worse. If, as Amy wanted, they financed her from the time she left Crockers until she could make a living as a commercial pilot, there was the distinct possibility of losing another daughter. On the other hand if they refused, they knew that she would become so frustrated, especially now that Hans had thrown her over, that there was every likelihood of a repeat of the nervous breakdown she had suffered whilst working for Frank Hall. It says a lot for the courage and wisdom of her parents that they went along with Amy's wishes.

Amy's flying career could never have succeeded were it not for their financial backing, which enabled her to leave Crockers at the end of the first week in September. From there on, the whole of her time was spent at Stag Lane, where she continued with her flying lessons. These now included instruction in simple aerobatic manoeuvres and solo cross-country flights. Whenever she was not flying, she was working in the hangar as Jack Humphreys's unofficial 'apprentice'. Under his watchful eye she graduated from sweeping the hangar floor to dismantling and cleaning greasy engine components, and finally to the more serious study of the intricacies of the design of the aero engine itself. At other times she would be found sitting warming herself in front of the fire in the clubhouse whilst she swotted for the exams that would eventually lead to her Ground Engineer's 'C' Licence.

The first really serious step towards a career in aviation came when Amy met the innovative engineer James Martin, later to be knighted for the invention and development of the life-saving aircraft ejector seat. He was the self-educated son of an Ulster farmer, a staunch Bible believing Christian and a man brimming with ideas, who was ultimately considered by many in the world of aviation to be a genius. As a young man he had raced cars at Brooklands, and it was from his involvement at this venue that he had first become interested in aeroplanes and their design. He subsequently formed the modest Martin's Aircraft Works with a small workforce at Acton, before moving to more suitable factory premises at Denham.

Jimmy Martin joined the club at Stag Lane early in 1928 and took flying lessons from Captain Valentine Baker, but after suffering a fractured skull whilst helping to erect a prefabricated steel building, he had been left deaf in one ear and as a result he had not gone on to achieve his 'A' Licence.

At the time he met Amy, he was concentrating on the design and construction of what he hoped would be a more modern replacement for the Cirrus Moth. The burly but gentle Irishman was not only an outstanding engineer but also a visionary who had the foresight and wisdom to concentrate on the monoplane, when most of the well-known British aircraft manufacturers were still involved with the outdated biplane configuration.

Amy discussing the route of her proposed cross-country flight with her colleagues at the London Aeroplane Club. (Martin–Baker Ltd)

Sir James Martin, the brilliant aircraft designer who encouraged Amy to make her flight to Australia. He eventually teamed up with Henry Valentine Baker to form Martin–Baker Ltd, a firm renowned for pioneering the design and production of the life-saving ejector seat.
(The Amy Johnson Estate)

James Martin also had the revolutionary idea of placing the engine behind the pilot. It meant using a 5-foot shaft from the engine to pass between the two side-by-side seats in the enclosed cockpit to the propeller, thus greatly improving forward vision for both pilot and pupil.* On the production side, Martin's ideas were for a simple tubular metal construction of the airframe, one which could be assembled and serviced easily and cheaply. Of course he would need someone to demonstrate such a machine, and it was at this point that Amy came into the picture. The plane's designer could see that there would be enhanced publicity value for his new brainchild if it were demonstrated by a lady test pilot.

Once the idea was put to Amy, she became wildly optimistic, seeing an opening for herself in aviation. Writing to her mother in September, she told her how she had been chosen to demonstrate this new and 'secret' aeroplane. She became a frequent visitor to the Denham factory, where the prototype monoplane was under construction, and where there would be discussions between James Martin and Valentine Baker on design changes. The two men eventually teamed up to form what is still today, the Martin-Baker Aircraft Company.

Exactly how the proposition changed from Amy being the plane's demonstrator to making a solo attempt on the UK–Australia record is rather uncertain. The most likely suggestion is that James Martin floated the idea to Amy and that she was then encouraged by Jack Humphreys. The Australian Bert Hinkler had been the first person to make such a flight, using a well-proven Avro Avian biplane, taking fifteen and a half days in February 1928.

---

* James Martin and Percy Waterman Pitt lodged a joint patent application in March 1931 for the positioning of the engine and the propeller shaft in this novel way. It is worth noting that the American Bell Airacobra P-39 Interceptor aircraft used a similar engine layout in the Second World War.

The two flying aristocrats already mentioned had soloed in leisurely fashion to and from South Africa, but no woman had yet attempted the 11,000-mile flight to Australia alone.

Amy was immediately taken with the idea, and was encouraged not only to make the flight but to better Hinkler's record into the bargain. What was not taken into account was the feasibility of doing it in an unproven prototype aeroplane which was still many months away from its test-flying stage. Amy had worked out that the ideal time to make the attempt on Hinkler's record was the first week in May of the following year, just over eight months away. However, there were two important factors which she did not take into account: first, James Martin was a perfectionist, a man who wanted everything exactly right, which meant that targets were almost never met; secondly, the Western world was about to feel the effects of the Wall Street crash, which was about to bring financial ruin to many businesses.

Although Amy had gained only limited experience of advertising when she worked for Morison's Agency in Hull, she had enough common sense to know that good publicity helps to sell a product. Quite fortuitously, a reporter from the *Evening News* visited Stag Lane in January 1930 to interview one of the pilots who was about to make a flight to India. Like all good newspapermen, he had an eye and an ear for the unusual story, and when he was asked if he had come to interview the young lady engineer (she had gained her ground engineer's 'C' Licence in November), who was also a pilot and one who intended to fly solo to Australia, he made an immediate beeline for the club's hangar.

Amy had yet to learn that what is said to a newspaper reporter and what is actually printed rarely coincide. When told that she was wanted for an interview, she turned to Jack and asked him what she should do. His reply was, 'Go for it. Make the most of your chance for some free publicity.' Questions were fired at her in quick succession. What were her aims in aviation? Where did she come from? How old was she? What type of aircraft was she hoping to fly in? How much did she earn? So it went on.

When Amy read the newspaper that evening there was an article stating that a young girl was about to fly a secret aeroplane alone to Australia, and that she was not only a qualified pilot, but also the first woman 'air engineer'.*

The article then went on to report that she was a twenty-two-year-old

---

* Amy was not the first, but she was now the only woman to hold a GE licence since Lady Heath had allowed her U S qualification to lapse.

Amy astride the cowling of one of the Club's DH Moths at Stag Lane as she makes final adjustments to its Cirrus engine. (The Amy Johnson Estate)

blonde* who came from the Midlands, and that she made a comfortable living from the work she did on aircraft and their engines.

Amy may have cavilled at the inaccuracies in the report, but if she was ever going to make the flight to Australia in Martin's machine – or any other – she needed financial backing, and to attract it she had to be newsworthy. Before long a string of photographers from Fleet Street and a crew of film cameramen arrived at the club, and Amy was filmed making a circuit of the airfield in one of the club's machines for the cinema newsreels. Photographs showing her in overalls, sitting astride the engine cowling of a Cirrus Moth with a spanner in her hand, began to appear in every London newspaper. No doubt there was some resentment amongst club members at the publicity she was getting, because many of them knew that she was far from being a natural pilot. It must also have placed a considerable strain upon Amy, as she now had to live up to the public's expectation of her, while knowing that the Martin prototype was far from completion.

There was an element of truth in the newspaper's statement that Amy earned a living as an engineer, inasmuch as she had recently been earning some spare cash by servicing an aircraft belonging to one of the members at the club, but it would hardly have amounted to 'a comfortable living'. Her flying log shows that between late December and early January, she flew Major Nathan's Moth, presumably after she had serviced it, from Stag Lane to Plymouth, prior to it being shipped abroad. Further entries in her log show that she flew another of his machines to and from Hamble on more than one occasion. This was not much to speak about, but it did show that Amy, albeit unwittingly, was taking one small step along the way towards what was then described as female emancipation. Some women protested and chained themselves to the railings outside Downing Street, others did it rather less obviously.

During all the time that Amy was engrossed in her new career, she never stopped caring about her two younger sisters, Mollie and Betty. She took time to write to them and was sensitive enough to realise that Irene's tragic and unnecessary death was liable to leave an indelible and damaging mark upon their young lives, particularly the impressionable eleven-year-old Betty. In a letter written in January 1930, she showed remarkable affection for her youngest sibling, and then went on to reprove her gently for her propensity

---

* Amy might not have given her true age, for one can see where the entry for her date of birth in her flying log has been altered to make the figure 3 of 1903, neatly into an '8'. Also about this time she had begun to streak and highlight her hair, and one wonders if it was a case of Hans's barbed jibe that 'gentlemen prefer blondes', which had prompted her to do this.

for daydreaming rather than getting on with her school studies. She wrote, 'You do love to sit in front of the fire reading those girls' novels. What does it matter if the heroine climbs out of the dormitory window. Better to grapple with your algebra and arithmetic.' She then concluded with an encouraging, 'You can do it!'

At the beginning of March Amy discovered that James Martin's project had collapsed. In spite of the successful completion of most of the aircraft's structure and the fact that the engine had been tested in its unconventional layout, a lack of money brought the work at the Denham factory to a complete standstill. She became desperate, for there were now several constraints bearing hard upon her. Not only had she not yet found a financial backer for the flight, but she now had the problem of finding a suitable aeroplane.

Added to these two major setbacks was the realisation that the allowance her father was giving her was not without limit. She had given him to understand that as soon as she had qualified as a pilot and a ground engineer, she would be self-supporting. And then, as if to make matters worse, there was friction between her and the club's Chief Flying Instructor, Major Travers. Some of the members were saying that she was flouting the club's rules by visiting the hangars, and Amy saw this as a pretext to prevent her from adding to her engineering knowledge. She also suspected, rightly or wrongly, that Travers was amongst those who objected; it was rumoured that he himself wanted to obtain his ground engineer licences and Jack Humphreys was unwilling to assist him in doing so.

A formal complaint was made to the club secretary, Harold Perrin, and Jack Humphreys was called for an interview. He defended Amy's presence in the hangar, pointing out that not only was she doing a good job, but she was also acting in an unpaid capacity. He then told Perrin that if she were forbidden to work there, he and the two other members of his staff would resign. The outcome was that Amy continued with her work experience unhindered, thus enabling her to obtain her other ground engineer's ticket, an 'A' Licence which qualified her to carry out the inspection of airframes as well as engines.

The dispute did, however, increase the animosity which seemed to have existed between Amy and Travers from the moment they had first met. At that time Amy was still traumatised from the humiliation and rejection that she had suffered from Hans Arregger, and whenever Travers reproved her for her landings, no matter how gently, she took it personally. So in Travers she did not see someone trying to help her, but just another man trying to put her down.

Eva Fitzpatrick describes her father's relationship with Amy as follows:

> The faint dislike which H and Miss Johnson appear to have felt towards each other on their first meeting had increased sharply as the months went by and as she persisted in stating her intention of flying to Australia, pestering H for his approval of her plans and generally badgering him on the subject. It appeared that she considered him, with his pleasant, well-modulated English voice and cheerful countenance, to be a typical product of a sheltered South of England public school upbringing – a man who had never done a day's work or known hardship or difficulty in the course of his easy, pain-free life.
>
> He considered her to be a grim and unattractive young woman with a chip on her shoulder as thick as three cubic feet of best British Columbia lumber.
>
> She considered him to be 'against' her because she was from Yorkshire and was a woman, for she was thinking of herself and so took his rebuttal of her plans personally.

Amy now embarked upon a furious campaign of letter-writing, contacting some of the most influential people and organisations in aviation and industry that she could think of in order to obtain the financial backing she would require to make the flight to Australia (she estimated it to be as much as £1500). Fortunately, she was familiar with the names and roles of those who were most likely to help her, for whilst waiting for her initial flying lessons she had undertaken to do unpaid secretarial work for the Air League of the British Empire.

Her output of correspondence even exceeded that with which she had pursued her relationship with Hans, and she now hurried to the post-box, not with a single letter but with a clutch at a time. Her targets included the Lords Rothmere and Beaverbrook as well as organisations such as the United Empire Party, but whilst she received sympathetic replies, there was still no real support in the form of hard cash. Lesser individuals would have given up at this stage and settled for whatever aviation employment offers came their way, but not so the stubborn girl from Hull. After much family discussion, her father offered to support her to the tune of £500, but generous as it was, she knew it was barely enough to buy even a second-hand machine.

The breakthrough came after Amy attended a lecture at the Royal Aeronautical Society, where the Director of Civil Aviation, Sir Sefton Brancker, gave an impassioned speech urging the development of the nation's air transport and communication routes to the Empire. Nothing could have suited Amy's purposes better. She sat down and wrote to him immediately in a challenging but respectful manner, stating that if she, a woman, could be seen to succeed in flying alone to such a distant part of the Empire as Australia, then it would encourage the public to have confidence in the possibility of mass air travel. In her excitement, however, she forgot to sign her name. Fortunately an article had appeared in the aviation journal *Flight* under the title 'Airisms from the Four Winds', in which she was mentioned. Brancker must have been very impressed with her letter, as he took the trouble to trace her address through this article, and she received a reply telling her that he would approach the philanthropic oil magnate, Lord Wakefield, on her behalf.

Lord Wakefield considered Brancker's request with some degree of caution, and it was precisely at this moment that Travers's influence became pivotal. Eva Fitzpatrick explained:

> The war between them dragged on. H won the early battles: 'She can't go. I've told her again and again, it is quite out of the question,' he said. But in the spring of 1930, just when he hoped that he had heard the last of it, she brought up some heavy artillery in the form of Lord Wakefield.
>
> Lord Wakefield, as reasonable as he was generous, had promised her both financial and material support for her proposed venture, support without which she could not afford to go, provided that the Director of Civil Aviation, Sir Sefton Brancker, thought that she was able to undertake such a flight.
>
> Brancker, loyal friend from the old Blackburn days in Greece, lost no time in getting in touch with H and sounding him out about Miss Amy Johnson.
>
> They talked long.
>
> H gave his opinion – that it was a brave idea but that she was unfit for such a journey. Her navigation was non-existent – she would lose her way – she relied too much on luck in her flying – she had no idea what such a journey entailed.
>
> Of course, if Brancker was prepared to assume full responsibility for Miss Johnson then that was another matter. He, H. G. Travers, would not do so. As a matter of fact it was a considerable relief, he said, to have the wretched business off his hands.
>
> Sir Sefton lunched with Lord Wakefield and put the case.
>
> Wakefield did not, of course, miss the point and the matter, which was to have been decided that day, remained unsettled for a further twenty-four hours. H had been badly mistaken in thinking that he had handed over responsibility to Brancker. Lord Wakefield had been counting on H's approval as a matter of course and it had not been forthcoming.
>
> Now Lord Wakefield wanted to see H alone, without Brancker, and he

asked him to lunch the next day. As a result of this summons H had twenty-four hours to consider and to decide Miss Johnson's future for that no longer lay with Sir Sefton but, via Lord Wakefield, with him . . .

No wonder he paced the room, up and down, up and down, three steps into the bay, three steps back, irresolute, worried, trapped. 'She doesn't realise the importance of such a flight . . . No woman has made such a flight before . . . The Press will fasten on to it . . . If it is a failure (and she's unlikely to succeed) it will be such a bad advertisement for civil flying . . . She seems to have no idea of the importance of what she is contemplating . . .'

All this was new to me. In my ignorance I had thought that people flew back and forth to Australia as a matter of course. 'Uncle Jim [Travers' brother] *always* flew to Australia,' I asserted with the confidence of the hopelessly wrong. (He had of course done nothing of the kind.)

H told me a little about mileages and fuel stops and oceans and weather. 'A few men have gone,' he said, 'a very few. And no women.'

He asked again: 'What do you think Eva?' I asked if it was a good thing to do. He said, 'Very good, if she ever gets there . . .' I said that she ought to go. 'You've made up my mind for me,' he said.

We then played our game of chess.

When I next spoke to my father on the subject it was some days after his lunch with Lord Wakefield. He described the scene, in the big room looking out on the London River – London River, the hub and the heart of the Universe. Beside that same river, from whence the Travers ships had sailed to the four corners of the earth, Miss Johnson's flight began at last to fall into the scheme of things and before they parted H told Lord Wakefield that, if she would undertake to learn some navigation, he would let her go.

Chapter Eight

Halfway Across the World

ood morning, Miss Johnson, please do take a seat.'

A slightly nervous Amy sat down on one side of the large mahogany desk and faced the doughty philanthropist, Lord Wakefield, a small bald-headed man with darting shrewd eyes set beneath heavy eyebrows. He wore a starched high wing-collar with a black tie above a white shirt and she noticed a gold watch-chain draped rather opulently from his waistcoat pocket across his portly midriff. Amy was only mildly optimistic for a favourable outcome of their meeting, especially now that her self-imposed deadline for departure was less than three weeks away.

Happily, Lord Wakefield announced that he was prepared to provide 50 per cent of the cost of a suitable aeroplane, providing her father was willing to put up the remaining half. In addition, he would provide the oil she required, as well as persuading petrol companies to provide fuel at the stops along her intended route. For his part, he hoped that her success would be good publicity for Wakefield Castrol Oils. But he was adamant that she should enrol immediately for a course of navigation at the Royal Aero Club.

A jubilant Amy set out to purchase the necessary aeroplane, a second-hand DH 60G Gipsy Moth, registration G-AAAH which had just come onto the market. Its owner was Captain Walter L. Hope, of Air Taxis Ltd, a small private company which operated from Stag Lane. The red and silver machine had been specially built for this charter company as a long-range single-seater and was first registered in 1928. It was fitted with the powerful 100 horsepower Gipsy 1 engine, giving it a top speed of 98 m.p.h. and had been used extensively in Europe and tropical Africa, clocking up 35,000 miles. Amy purchased it for £600 on 30 April.

Although the plane had only completed fifty flying hours since receiving a top overhaul it went into the de Havilland workshop to have a bucket-seat installed to take a parachute (an item of equipment Amy's parents insisted upon), and for a complete respray to give it a dark bottle green fuselage with silvered wings and tailplane. It was then taken over by Jack Humphreys, who removed the engine for cleaning and a thorough inspection; at the same time incorporating some recently introduced minor modifications. Amy wanted to personalise her new acquisition and so it was given the name *Jason*, painted in silver on each side of the engine cowling. This was not, as many thought, in recognition of the mythological Greek hero who went in quest of the Golden Fleece, but less romantically, the registered trademark of her father's fishery business.

Relations between Amy and Travers immediately began to improve as she responded to his guidance on navigation. He was never one to leave pupils 'hacking around the airfield' as he called it; he much preferred to take them off on short cross-country flights. Past pupils, amongst them the well-known pilots Francis Chichester and Jean Batten, never forgot his final instruction before sending them off alone on such flights. 'Don't forget the three Cs,' he would say. 'Clock, compass and common sense.'[1]

Amy's first solo cross-country flight was not a success, as she ended up completely lost and was forced to make a landing at Stony Stratford in order to find her whereabouts. She subsequently made several short excursions away from the airfield, following triangular courses to such places as Dunmow, Duxford, Reading and Brooklands, and then the Nathan flights to Hamble. However, right up until her departure for Australia she never flew solo further than from Stag Lane to Hedon (Hull) and back, and certainly never made a sea crossing.

Amy was not entirely ignorant of the academic side of navigation, for earlier in the year she had enrolled for her 2nd Class Navigator's Licence at the de Havilland Technical School at Stag Lane. It was whilst she was on this course with other members of the LAC that she became acquainted with Lady Bailey, who was also attempting to master the intricacies of the Morse code, so necessary for wireless telegraphy, and other subjects such as map-reading, meteorology, dead-reckoning (an unfortunate term), magnetism and compasses. As a result the two women became very friendly.

Since the formation of the Guild of Air Pilots and Air Navigators in 1929, a licence in navigation had become necessary for the acquistion of a commercial pilot's 'B' Licence, something which Amy had to acquire if she was ever to achieve her ambition of a career in aviation. However, she soon came to realise that she had taken on too much. The long hours involved in studying for her ground engineer licences, together with the time spent in correspondence seeking financial backing for her flight to Australia, had left her with no option but to abandon the course on navigation at the school. Her father, who was primarily concerned for her health, had advised her to delay the Australia flight for a year so that she could complete these studies in a more leisurely fashion, but Amy was determined to go ahead with the flight by the first week in May 1930, or not at all. So, with just three weeks to departure, she was having to undergo an intensive course of navigation at the Royal Aero Club.

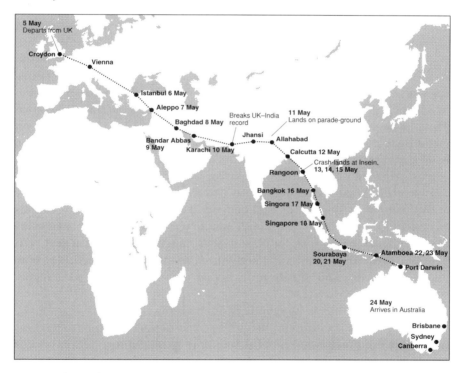

Apart from the extra training she had to absorb in so short a time, she also had to submit to Wakefield Oils her intended route and the places at which she expected to put down, so that oil and fuel supplies could be arranged. Bert Hinkler's route in 1928 had gone south across France to Rome and thence across the Mediterranean via Malta to Tobruk. The little Australian then flew eastwards along the North African coast to Ramleh in Palestine and then down to Basra on the Persian Gulf. Amy's plan was to chop 600 miles off Hinkler's route by flying directly across Europe and over the Balkans to Constantinople (Istanbul), thereby avoiding a sea crossing. She then intended to fly south to Baghdad and to link up with Hinkler's route at the Gulf. Her final route, therefore, would be from Croydon to Vienna, Constantinople, Baghdad, Bandar Abbas, Karachi, Allahabad, Calcutta, Rangoon, Bangkok, Singapore, Sourabaya, Atamboea and across the Timor Sea to Darwin. She estimated that with a fuel capacity of 80 gallons, sufficient for thirteen hours in the air at 90 m.p.h., she could reach each of her stopping places before sunset. There was just one snag: she needed visas for each of the countries over which she would be flying in the Balkans, something that could often take more than the three weeks she had available. However, with the assistance of Sir Sefton Brancker they were speedily acquired.

The amiable Captain Wally Hope, from whom Amy had bought her aircraft, was well respected in aviation circles as an experienced pilot, and since he had first-hand knowledge of flying the machine under tropical conditions it seemed only natural for Amy to seek his advice. He suggested she use roller strip maps instead of map folders, as they were much easier to handle in a draughty open cockpit, and warned her to wear a sun helmet. Then there were the less obvious things, such as a first-aid kit, quinine for malarial protection, emergency rations and a portable stove. In case she had to make a forced landing in an unfriendly area he recommended that she carry a revolver, especially being a young woman on her own. It was good advice of which she took careful note.

In a letter to her mother around this time, Amy sought to allay her worries by telling her, 'I'm taking every precaution, you may be sure. I am getting heaps of things for nothing – watches, clocks, engine parts, new wheels, etc. It's great fun.' From such comments one feels bound to sympathise with Travers's view that Amy did not really appreciate what she was taking on.

Just before sunrise at 4 a.m. on Monday, 5 May 1930, Amy was woken by Will Johnson, who was with her at Croydon's Aerodrome Hotel. 'How did you sleep, Amy?' he asked.

Propping herself up on one elbow, his bleary-eyed daughter replied, 'Awful, I couldn't sleep for traffic noise.' (The hotel was situated alongside the busy Purley Way.) She dressed quickly and took breakfast before walking the short distance to where her aircraft stood waiting outside the open hangar doors. Throughout the night it had been worked on by Jack Humphreys and James Martin, refuelling, greasing and oiling, and trying on the spare propeller which was now strapped to the centre section struts on the port side of the fuselage. It would slow the biplane by causing increased

Amy pictured at Croydon Airport on the morning of her departure for Australia. Note the spare propeller tied to the upper wing struts. (Hull Daily Mail)

Will Johnson, James Martin, Jack Humphreys, Flying Instructors, Herbert Travers, F.R. Matthews and Gordon Store bid Amy farewell on the Monday morning of 5 May 1930. The only woman amongst the small group is Winifred Irving (second from the right.) (Martin-Baker Ltd)

drag, but Amy knew that in the event of a mishandled landing the propeller was usually the first thing to be damaged. Apart from a customs official, two airport workers, a reporter and press photographer, the only people to see her off on the cold, grey morning were her father, James Martin, Jack Humphreys, Winifred Irving and three of the Stag Lane flying instructors – Travers, Matthews and a newcomer to the club, Gordon Store.

After a kiss and a hug from her father and a round of handshakes and farewells, she climbed into the rear cockpit and waited for someone to swing the propeller. As she was fastening her seat and parachute belts she began to smell petrol, and then noticed that fuel was dripping from one of the main fuel supply connections. She was advised to go back to the hotel and try to rest for a few more hours until the fault was corrected. As a result it was 7.45 before she was back in her machine and ready to depart. She had never flown a heavily laden Moth before, and *Jason* was well over the manufacturer's recommended loading. Apart from the additional weight of propeller, spares, tools and all the paraphernalia needed for such a long flight, the machine was carrying an extra 61 gallons of fuel, 35 in the front cockpit and 26 in the rear locker behind her seat, in addition to the normal 19-gallon gravity tank in the centre section of the top wing.

On her first attempt at take-off the machine was reluctant to leave the ground and Amy was forced to abandon the run. She turned around and taxied back to try again. According to Sir James Martin he did a few quick calculations and told her to pull the stick back when she came to where he would be standing. He then paced out the distance and stood waiting for her. When she drew level she followed his instructions and the silvered wings began to lift the machine steadily into the sky. Then came a cheery wave from the young pilot as she set out on a south-easterly heading for Vienna. One cannot help but wonder if, as she took one last look down at that vast sprawl of London's suburbs, she said to herself, 'Well, Mr Arregger, I wonder what boring kind of a life you're living down there now.'

The neat, hedged fields of the Kent countryside slid lazily beneath the fabric-covered wings of the tiny aircraft as it made its way towards the English Channel and on to her first destination. The even purr from the open exhaust stubs of the Gipsy engine gave her a comforting reassurance, in spite of a slight headwind and the few spots of oil that settled from time to time on the small windscreen in front of her. These were easily removed by the occasional wipe with the back of her glove, but were still a nuisance. What was more annoying was that she needed to pump petrol from the fuselage tanks to the top wing tank by means of a hand pump. It was a new and unwelcome experience, for apart from the flight from Stag Lane to

Croydon with Jack Humphreys on the Sunday before she left, she had only made one brief test flight in G-AAAH. She soon discovered that her biceps were about to be well exercised, as she needed to pump forty times to get one gallon into the gravity tank – no joke when she had to pump 50 gallons a day!

After ten hours in the air she was circling the large modern aerodrome at Aspern and making her first landing of the journey. So far so good. Amy wrote:

> At Vienna the mechanics absolutely refused to let me do anything, and I danced hither and thither explaining what must be done, my head full of my last promise to my engineer tutor to do everything myself. The mechanics meanwhile endeavoured to amuse me by a story of their last lady pilot visitor who changed into overalls and mounted a ladder, insisting on looking over the machine herself. Now if there is one thing I cannot stand, it is the idea of being laughed at behind my back, so I left them to it, and went off to see about my passport and finding somewhere to sleep, as there was not time to go into Vienna.[2]

At 4 a.m. Amy was ready to depart and was surprised to find that aerodrome staff were on hand to assist her to start the Moth's engine. It sounded decidedly rough and she found that when switching from one magneto to another there was a serious drop in engine revolutions. Diagnosing a fouled-up sparking plug, she took out her tool kit and removed and cleaned the plugs. It cured the problem. After this incident she made a habit of cleaning the plugs herself at the end of each day's flying, since it was not a chore with which she wished to be delayed by in the mornings.

Chocks away! First stop Vienna. (Martin-Baker Ltd)

There can be little doubt that a significant factor in Amy's success on her record-breaking flights was that, having been 'one of the boys' when she worked on the aircraft in the hangar at Stag Lane, she had a natural rapport with the aerodrome ground staff that she met along the way. Moreover, she gained the respect of the mechanics who assisted her, for they came to realise that she had first-hand technical knowledge of the machine she flew. It was something they did not expect in a lady pilot!

The 800-mile flight to Constantinople was marked by severe rainstorms, with white pellets of rain bouncing back along the fuselage and into the cockpit and soaking the anxious pilot. To make matters worse, the petrol pump developed a leak, causing fuel to squirt into her face, so that she was forced to pump the handle whilst leaning with her head over the side of the cockpit. She arrived at the San Stefano aerodrome just before sunset, but was so delayed by the customs and passport-control officials that by the time she got back to service her aircraft it was dark. Fortunately, there was a friendly car owner on hand who beamed his headlights onto *Jason* so that she was able to replenish the engine oil and clean the filters and plugs.

When it came to refuelling at this primitive 'airport' she managed to rally the assistance of the Turkish authorities. It was a slow and tedious job for it meant that the contents of each can had to be poured carefully into the three tanks by hand, using a chamois leather as a filter to prevent water or dirt getting in, which could easily clog one of the carburettor's jets and cause an engine failure. Once the task was completed, she used sign language to request that her machine be placed in the hangar overnight. In their enthusiasm to help, and before she was able to restrain them, several hefty Turks lifted the tail up to push the aircraft towards the hangar, causing it to nose over onto its propeller hub. Fortunately, Amy's good engineering training had prompted her take the precaution of positioning the propeller blade horizontally, or else it might well have been damaged and pushed out of true.

So far Amy's flight had attracted very little attention from the British press, since few people imagined that she would ever accomplish what she had set out to do. The only surprise waiting for her at Constantinople was a cablegram from Hans wishing her every success. We do not know how she reacted to this unexpected congratulation; she probably gave it little thought – perhaps a wry smile, nothing more. However, it did show that her former lover was studying her progress, no doubt with a great deal of astonishment.

In spite of letters of introduction from Sir Sefton Brancker which were supposed to smooth her way (her Turkish flying permit had not arrived when she left the UK) and to counter any queries on her flying over Turkish airspace, the authorities at San Stefano would not release her aircraft until they had confirmation from one of their own government officials. Much to Amy's annoyance, there was thus a lengthy delay in the morning and she was forced to abandon her plan of making Baghdad in one stage before nightfall. It now meant she must fly the shorter 600-mile route to Aleppo in Syria. When she did get permission to take her aircraft from the hangar and ran the engine up on the ground she discovered that the oil pressure was only reading 10 pounds per square inch instead of the normal 30–40. At the same time she smelt the pungent odour of petrol, signifying yet another leak. The low pressure she traced and rectified herself by freeing a sticking plunger in the oil-release mechanism, and she used her feminine charm and linguistic skill to gain the assistance of a French mechanic, who happened to be working nearby on one of his own nation's machines, to fix the petrol leak.

It was 10 a.m. before the wheels of the overladen Moth finally parted company with Turkish soil and the machine headed towards the barrier of the Taurus mountain range, which Amy must cross if she were to make the Mouslimie aerodrome that day. She was relieved to notice that there was no vibration coming from the propeller and concluded that no real damage had occurred from the mishap of the day before. A height of 11,000 feet was needed to clear the mountains, which she could now see towering in the distance ahead, but the engine was not happy as it approached this altitude. There were now three choices: turn back, keep going and hope she could gain enough height to scrape over a lower part of the range, or descend and follow the railway line which snaked through narrow, twisting gorges and sometimes disappeared disconcertingly into tunnels. She decided on the last. What she might not have been aware of was the dangerous downcurrent that is often encountered when flying into narrow valleys or ravines.

Nevertheless, all went well until Amy was flying along a twisting valley which suddenly became obscured by a bank of low-lying cloud. On each side of her were sheer rock faces, so there was no possibility of turning back. She pushed the stick forward as she entered the cloud in a desperate attempt to descend below it, and emerged to regain sight of the railway line with one wing of her machine perilously close to a towering wall of rock. She felt a cold shiver run down her spine. Amy was not a particularly religious woman, but she admitted some time later that she was conscious of God's providential hand upon her on many occasions. Her mother was a devout and committed Christian and one could believe that this small woman was much in prayer throughout the time of her daughter's flight.

Amy landed into the shimmering heat of the northern Syrian desert in the late afternoon of 7 May to be greeted by the smell of oil on hot sand. As soon as she climbed down from her faithful *Jason* she immediately began to peel off the heavy outer garment of her Sidcot flying suit and made herself more comfortable in an open-necked shirt, riding breeches and boots, before tackling the work on her aircraft. She always maintained that the hardest part of the flight was the servicing that she had to do on the ground. However, refuelling at Aleppo was not the chore it had been the previous evening, for here the French air force were on hand to fill the tanks from a modern garage-type pump. Then with the help of some French mechanics she carried out the necessary service to her engine, changing oil and plugs and inspecting the suspect plunger. It was not every day that an attractive young lady pilot called into the aerodrome, and she soon found herself the centre of attraction, as some of the officers asked her to pose for photographs.

It was at Aleppo that she gained her first impressions of the mystical East, and the chance to see real Bedouins and sheikhs, their dark brown semitic eyes meeting her nordic blue in a mutual inquisitiveness, so different from the sanitised versions she had seen on the silver screen as a girl. The sights and sounds of this strange land intrigued the young woman from Hull. That evening the rustle of the palm leaves in the breeze and the twinkling clarity of the stars at night, something unknown in northern Europe, bewitched her, whilst the soft footsteps of the sentry padding up and down outside her quarters in the barracks soothed her into a gentle sleep. She had always longed for adventure, now at last she had achieved her ambition. She slept soundly that night.

Early-morning starts were the norm for each stage of Amy's flight, since not only did they mean she spent less time in the heat of the day, but more importantly they were also a precaution against any unforeseen difficulties that might arise. This particular Thursday morning was no exception as she commenced her 500-mile flight to Baghdad. After crossing the glistening thread of the Euphrates, which snaked south-eastwards beneath her wings, she headed out on her compass course across a featureless and trackless waste. There was little of interest to be seen except a reddish-brown desert, which faded into the smudge of a grey horizon.

In order to avoid the worst of the heat and the turbulence that always accompanied it she climbed to 7000 feet, into cooler air. Then directly in front of her she saw the ominous sign of an approaching sandstorm. As she entered it *Jason* was thrown violently around the sky, its engine cutting in and out, and quite suddenly she was descending rapidly and uncontrollably.

Sand particles stung her face and even found their way beneath her goggles as she struggled to make a forced landing.

According to Amy's own account she landed at the incredibly high speed of 110 m.p.h. hoping against hope that there were no ditches or boulders in her path. She switched off the engine, clambered out and pulled the tail of the machine around so that the nose was facing into the wind. Fearing that the rocking biplane would be blown backwards, she took pieces of her luggage from the front compartment and used them as wedges under the wheels. Fastening the canvas engine cover to keep sand and dust out of the carburettor, which normally took only a few minutes, now became a frantic battle against the wind. No sooner had she tied one side of the cover and raced around to other side than the first side would be whipped in the air and torn from its fastenings. She then climbed up onto the engine cowling and tied her handerkerchief over the air-vent hole of the petrol tank to prevent sand from entering. The next three hours were spent with her back to the wind sitting on the tail of the machine to keep it down.

The dust storm passed and miraculously the engine fired first time, so that Amy was able to take off and make Baghdad within the next hour. As the machine touched down it sank onto its lower port mainplane and swung around as the left-hand undercarriage radius rod snapped, obviously weakened by the heavy landing in the desert.

Fortunately, the fitters in the workshops at the nearby station of RAF Hinaidi were able to make a completely new strut overnight and replace it before the following morning. In the meantime, Amy managed to delegate the refuelling and servicing of *Jason* to the mechanics of Imperial Airways (forerunner of British Airways), and took a sight-seeing tour around the city before dining at the rooftop restaurant of a nearby hotel, overlooking the Tigris. She had a great deal of modesty, as can be seen in the paper she gave before the Society of Engineers in October 1930, where she said, 'I knew that proper men mechanics would do the job much better than I could, so I had no compunction in leaving it to them whenever I could.' This is small comfort perhaps for those who would seek to portray Amy Johnson as an incipient feminist.

Even before Amy left Baghdad, the newspapers at home were beginning to scent a story, for she was within striking distance of Hinkler's time from the UK to India. James Martin, acting upon Amy's instructions, had visited the offices of the *Daily Express* and the *Daily Mail* on the day of her departure to sell her story, but had been turned down flat. Now he was being offered £1000 by the *News of the World*, a sum he considered too low. He was finally offered £10,000 by the *Daily Mail*, which he accepted on her behalf.

When Amy took off for Bandar Abbas on the eastern shore of the Persian Gulf on the fifth day, she was wearing borrowed cotton shorts, a shirt and sun helmet, for she knew that the heat on this stretch of the journey would be almost unbearable. When Sir Alan Cobham landed at Baghdad in June 1926 on his way to Australia, he recalled that the heat brought him 'practically to a state of collapse'.[3] Temperatures of 110–118° F (43–48°C) in the *shade* were not uncommon, enough to immobilise the fittest of men.

In order to avoid the worst of these conditions, Amy took her machine up to 10,000 feet. Not only did this make for a cooler journey, but it also avoided the very real danger of being fired upon by trigger-happy Marsh Arabs. These armed tribesmen, who were renowned for their accuracy with a rifle, regarded all aeroplanes as part of the RAF, an enemy occupying their land.* Amy maintained her safe height, following the muddy brown snake of the Tigris down to the northern shore of the Gulf and on to her next scheduled stop. She recorded:

> At Bandar Abbas I was not expected; the aerodrome was not marked in any way, nor was there any indication of the wind. Presuming there was a sea breeze I landed towards the Gulf in the only available large space, which proved to be the aerodrome. I landed fast, as usual, and rather heavily, also as usual, in the terrific heat I was now flying in, and to my horror, the left wing drooped and trailed the ground. This time it was the bolt securing the top of the new strut which had sheared, and I could not see any hope of help in such a Godforsaken place as Bandar Abbas.

The high-speed landings which she referred to were due to the thinner, lower-density air found in the tropics, which causes an aircraft to sink more rapidly than normal, a condition that would require increased speed in order to make a safe landing. Similarly, she found that her machine required a considerably longer take-off run than usual to become airborne. The sheared bolt might well have been due to the replacement not being of the necessary high-tensile quality for such a highly stressed component.

Had Amy had less gritty determination she might well have given up at this setback. Instead she dragged herself towards the large house that she had seen from the air to be greeted by the Consul, who told her that the airport was no longer in regular use and, moreover, he had not been warned of her arrival. As soon as his wife met her, she persuaded her to lie down in the cool of an upstairs bedroom for she was suffering from a splitting headache.

* After the fall of Germany and its Ottoman ally in 1918, several of their former territories were handed over to the Allies, with the result that Iraq, Palestine, Jordan and a large part of Persia had been mandated to the UK under a system adopted by the League of Nations in 1920.

Amy slept soundly for a few hours and when she awoke she was surprised to find it was dark. Her immediate thought was for *Jason* and the repair that would be needed if she was to continue her flight in the morning. Making her way outside into the cool air, she was amazed to find her aircraft standing perfectly upright and normal. Had she dreamed the mishap, she wondered. She soon discovered, however, that the airport's 'man Friday', a young man named David, had matched the broken bolt with one that he had in a collection of spare parts that he had acquired when RAF machines had been stationed there. What was more, he was prepared to assist her in servicing the engine as soon as she had dined with the Consul and his wife.

It appears that the Persian customs officials at Bandar Abbas ran a lucrative system of quite unlawful restrictions on visiting foreign pilots, which forbade them leaving the country unless they greased the officials' palms with cash or kind. Other long-distance aviators, such as Cobham in 1926 and Mollison five years later, had experienced a similar kind of blackmail. Amy was confronted by white-coated customs officials at the Consul's home, demanding to see a health certificate showing that she had been vaccinated against a long list of tropical diseases. She did not possess such a thing, nor had she been required to show one in any other country. She had been warned of these tricks before leaving the UK and had been advised to be courteous but to stand firm and not give way. After much flattery and polite argument, which lasted for almost two hours, she was actually allowed to sign her own health certificate and proceed with her flight.

She went out to her machine at 10.30 p.m. after a delayed evening meal, and by the light of her torch and a rising moon she was able, with the assistance of the helpful David, to complete the work of refuelling and servicing. She recalled that she was not in the least bit tired by the time they had finished the work at 2.30 a.m., and attributed this to the refreshing cool night air. One wonders, however, if it was not because of the excitement that she must now have felt, for if she made Karachi that day she would have knocked two complete days off Hinkler's UK–India record. At 4 a.m. she was wide awake and preparing to leave for the 700-mile flight which lay before her.

According to Amy, this stage of her flight, as she followed the mountainous Persian coastline eastwards, became 'very worrying . . . partly because my engine was again "missing" on the right-hand magneto, and I could not imagine what was wrong, as I had been so extra careful the night before.' Her other concern, one which bothered her increasingly since flying into tropical temperatures, 'was the air vent in the front petrol tank, which was a small pipe leading into the front cockpit.'

I had no gauge for the front or rear tank, and the only way to tell they were full was to put a finger inside, as it was not possible to see the petrol level. Often the tanks overflowed, and even when the front tank was not full, the vent overflowed as it was some distance from the top of the tank. The result was that the whole of my kit in the front cockpit became saturated with petrol, and when I retired at night I had to do so in petrol-soaked pyjamas . . .

The air vent problem was exacerbated as the sun rose higher and bore down onto the fuel tanks, causing them to exude sickening fumes into the rear cockpit in which she was now being well and truly microwaved.

With one anxious eye on the oil-pressure gauge and the other on the lookout for a smooth beach on which to make a forced landing if necessary, Amy pressed on, passing the small fishing ports of Gwadar and Pasni which lay along the Baluchistan coastline. Apart from coping with the jolts and bumps of air turbulence, severe enough in a modern 747 let alone in a small biplane, an additional stress was that the Gipsy Moth, unlike today's aircraft which can be flown 'hands off' by using auto-pilot, required Amy's full time concentration to hold it on an accurate compass course. This in itself must have played a large part in the fatigue she felt during her flight.

Karachi was known as 'the gateway to the East' and as Amy throttled back to descend to the Drigh Road aerodrome, she would have seen the huge 201-foot high hangar (reputed to be the tallest building in India), built to house airships, such as the ill-fated R101, which were intended to link the UK with the far parts of the British Empire. As she landed there was a great deal of excitement for the 'young London typist' had beaten the men at their own game – she had reduced Hinkler's time to India by two whole days. The international press were calling her the feminine equal of Lindbergh, for she now had a real chance to beat Hinkler's time of fifteen and a half days to Australia.

Amy's immediate concern upon landing was to investigate the cause of the engine misfire which had given her so much concern. It soon became apparent when she noticed a round black bulge on the outside of the engine cowling, and upon further inspection saw that one of the plugs had been shorting out as its terminal touched the cover when closed. At Bandar Abbas she had passed a new plug to her 'man Friday' David, stressing the importance of making sure it had a new washer underneath it, but he had evidently failed to remove the old washer which was still in the recess of the cylinder head. The result was that two washers now remained under the plug, causing its increased length to make contact with the cowling – a small error perhaps, but one which could have resulted in a forced landing in Baluchistan, not the most friendly of places for infidel foreigners. Writing in the Imperial Airways *Gazette* in 1932, A. B. Thompson, the company's station superintendent at Gwadar, described the Baluchis as 'an illiterate class of people, eking out a precarious livelihood by rearing and selling cattle, milk, dates, mats, coal, firewood and fish.' More to the point, as Alexander Frater recalled in his book *Beyond the Blue Horizon*, 'Gwadar lay at the heart of "ball chitty teritory", where strangers were advised to carry reward notices in Baluchi stitched to their coats, promising a £20 cash payment if the bearer was returned alive and without his testicles stuffed in his mouth.' One only hoped that the natives' illiteracy did not prevent them from understanding written instructions!

Once again Amy received unsolicited help from resident mechanics, this time Imperial Airways staff, who told her to get some sleep whilst they serviced her machine overnight. With her new celebrity status, she was driven away in royal fashion to spend the night, not in the Killarney Hotel where Imperial Airways' passengers usually stayed, but at Government House itself. She retired early, politely declining the invitation to attend a dinner party which had been proposed in her honour by pleading fatigue and a lack of suitable attire for the occasion. After dining in her room and getting a good night's sleep, she was up early for breakfast and back in the hangar before daybreak.

She had mentioned to the mechanics that her engine appeared to be running too rich, so they had changed the carburettor's jets. Now, sitting in the cockpit running up her engine in the dark, she was pleased to note that the bunsen-blue flame which pulsated from each of the four open exhaust stubs indicated that they had corrected the carburation to her satisfaction. Moreover, they had gone through all her tools, replacing those that were missing and cleaning the spare plugs she was carrying, before wrapping them in fresh petrol rags.

Amy taxied out across the dusty airfield and into a welcoming sunrise, ruddered her machine into the wind with a judicious blast of the throttle and was off again on the seventh day of her journey. It says a great deal about the intense interest that was now being shown in her progress that this time she was escorted for a short distance by the local de Havilland agent flying a Moth similar to her own, and a machine of the RAF. There was a brief exchange of hand salutes as Amy banked and turned eastwards to fly out over the grey-brown waste of the Sind Desert. *Jason*'s silvered wings glistened in the soft rays of early sunlight as it disappeared in the direction of Allahabad, some 900 miles away.

After flying into strong headwinds for almost ten hours, Amy estimated that she must be nearing her destination. She knew that Allahabad lay at the confluence of the Ganges and Jumna rivers, although she had been warned that rivers could be dried up at this time of the year, just before the monsoons. Seeing a large town alongside a solitary river she assumed it was her intended target, but as she circled it she failed to locate the airfield of Baumrauli shown on her map. Finally, she decided to land in an open space outside the town and make enquiries. However, as soon as she touched down she was met by a crowd of natives running towards her like a swarm of bees. They were not hostile, merely inquisitive, for in all probability they had never seen an aeroplane close up before. Unable to make herself understood she remained seated in the cockpit until at last an English-speaking man stepped forward from amongst the crowd.

'Where am I?' she asked.

'Jhansi,' he replied.

'How far is Allahabad?'

'Two hundred miles that way,' he said, pointing east.

Amy took off again, hoping to reach Allahabad before dusk, but after a short while in the air her second reserve tank began to run dry and she knew that it was too risky to chance making Allahabad with just the fuel she had left in the top gravity tank. Reluctantly, she decided to return to Jhansi, but eschewed another landing on the open space which she had just left. Circling the town once more she opted for what looked to be a suitable spot, not too isolated and with what she thought were a few houses nearby. The location she had chosen was in fact the parade ground of the 3rd and 8th Punjab Regiment, whilst the 'houses' were their barracks. The Gipsy Moth was not equipped with a braking system and as she closed the throttle and rolled to the end of her landing run she could see that she was heading straight for a post. Unable to stop, she struck the obstacle and came to a halt. She was horrified to see that she had damaged the leading edge of the wing.

The Punjabi officers and natives could have not been more helpful, for not only did they assist her with servicing the engine, they also arranged for the village carpenter to come out and repair the damaged wing. Fortunately, Amy's kit of spares included some fabric and dope with which the local village tailor was able to patch the torn area of the wing. Amy seemed to take the incident in her stride and later reminisced:

> The heat was terrific, but I landed in the best possible place – the Colonel's front yard practically – and servants came constantly to and fro with long cool drinks. As time wore on my kind friends made me lie down on a camp bed near the machine, and direct operations. This was

the nicest way that I had ever overhauled my engines. I liked it! Enough petrol was fetched in cars from the aerodrome 10 miles away to enable me to reach Allahabad.[4]

The next morning Amy made a stop-over at the central city of British India and the capital of the United Provinces. The sacred Ganges, alongside which the city of Allahabad stood, was the annual pilgrimage centre for all good Hindus, a place where they hoped their sins could be washed away. For Amy it meant little more than a dusty, arid airfield and a place to refuel. After spending more time than she would have liked removing and replacing the propeller in order to examine and clean the magneto points, a task found to be necessary to cure a misfire which had developed during the flight from Jhansi, she pressed on to Calcutta.

By the time her travel-stained Moth bumped down on the Dum Dum aerodrome* trailing a swirl of dust in its wake, she had made a total distance that day of 750 miles in stifling heat. Her name was now world news and a crowd was out to welcome her. Everyone was willing her on to achieve her goal of setting a new record to Australia, something that was still well within her grasp. Members of the local flying club volunteered to relieve her of the chore of servicing her machine, thus enabling her to spend time acquiring fresh clothes before resting.

Meanwhile, as she would be flying the next day over the mountainous and dense jungle areas of Burma, where a forced landing could be fatal, checks were made on the Gipsy Moth's compass for accuracy. The aircraft was placed on a circular concrete base on which were

Amy reading telegram cables of congratulation at Calcutta.
(Author's Collection)

* It was near here, in a small arms factory, that Captain Clay invented the hollow-nosed dum-dum bullet in 1898, one which expanded on impact to cause extensive flesh wounds.

marked true magnetic bearings. As the metal portions of an aircraft's structure affect the accuracy of its compass, it is necessary to correct these inaccuracies by placing small magnets inside the instrument. Eight readings were then taken at 45-degree intervals, adjustments made and the deviations then marked on a card attached to the front of the compass. This would enable Amy to make the necessary allowances on her course-setting.

Amy had no experience of flying in monsoon weather, but she knew enough about it from the folklore gleaned from other pilots to realise that the best way of avoiding the extreme turbulence which accompanied its convection currents, was to fly as low as possible. She also knew that the 'Burma rains' were about to break in mid-May – weather reports, where they existed, covered no more than local conditions; long range reports were still many years away. Apart from these hazards, she was flying without radio and in an open-cockpit machine which gave little protection from the elements.

Leaving the flat marshlands of the Ganges delta, Amy hugged the coastline of the Bay of Bengal, skirting an archipelago of small volcanic islands which were scattered like jewels in a turquoise sea. As she flew south towards the small fishing village of Akyab, she began to gain altitude in order to cross the 10,000-foot range of the Arakan Yoma, which towered in the distance. Its highest peaks were shrouded in dark, anvil-headed cumulonimbus clouds which were to be strictly avoided by frail biplanes. On the far side of this craggy range of forest-clad canyons lay the Irrawaddy, a river that would fix her position for the final descent to the racecourse at Rangoon.

As she threaded her way through a labyrinth of valleys, many of which were capped in forbidding clouds, she finally emerged to where she could see the gleam of a river beneath her. However, as she descended to follow it southwards, she ran into blinding, torrential rain. It fell with a ferocity she had not known before. So this was a monsoon. White pellets of water cascaded back off the propeller and into the cockpit and onto her goggles. She could barely see as she turned inland to pick up the railway line which would guide her to the Burmese capital.

With daylight fading fast and petrol running low, a very tired Amy looked for the racecourse, which according to her rain-sodden map lay alongside the railway line and on the north side of the city. All she could see was a small area of open ground, which might possibly be the racecourse. As she throttled back and made her approach, she realised much too late that the landing area was far too small, more ominously she could now see goal-posts ahead of her. There was little she could do – she was committed to a landing. As she recalled, 'There was not the slightest chance to take off again because of the high trees and buildings.'

The Gipsy Moth narrowly missed the posts and ran on towards a wire fence which was shielding a ditch. How she must have longed for brakes to arrest her progress. There was a splintering crash as the machine struck the fence and nosed over into the ditch. The port lower wing crumpled as it hit one of the fence posts, whilst at the same time it ripped the left tyre open and bent the oleo leg cowling of the undercarriage.

As she scrambled down from the cockpit of her wrecked machine there was an uncanny silence, broken only by the soft sound of sobbing as her tears mingled with the unremitting rain. Her immediate thought must have been, 'This is the end of my Australia flight.'

Chapter Nine

CALL ME JOHNNIE

The only consolation Amy could draw from crash landing where she did was that she had finished up on a playing field belonging to the Insein Engineering Institute. With a touch of humour she said, 'I was at any rate relieved to know that I was near an engineering school, and not a dancing academy.'

Fortunately, the head of the Institute was a kindly and sympatheteic man, who immediately arranged for the crumpled *Jason* to be dragged from the ditch and brought into the workshops. The next morning, in spite of it being examination time, he placed the whole of his works, staff and pupils, at Amy's disposal so that she could oversee the repair work. The damaged wing was removed and with the help of the Forestry Inspector and a team of his men, new wing ribs were made and replaced. They worked throughout the night and Amy recalls that several native servants were needed to swat and wave off an army of insects that made life difficult. The following morning someone remembered that a load of aeroplane fabric which had been surplus to wartime requirements had been sold off cheaply to local Burmese women who had subsequently made it into shirts for their husbands. As a result, as many shirts as could be found were torn up into strips and used to patch the wings before doping and painting.

In the meantime, the torn tyre was vulcanised and damaged parts of the undercarriage were repaired by the students under the supervision of the staff, whilst Amy replaced the shattered propeller with the spare she carried, and serviced the engine. The repaired machine, with its wings folded, was then towed the 12 miles to the Rangoon racecourse, proceeding at a snail's pace and stopping every fifteen minutes to allow the tyres to cool down in case the vulcanised cover gave way.

By the time they reached their destination it was almost sunset, which meant that Amy's departure was delayed until the morning. She had lost two whole days and now had but a slender hope of beating Hinkler's time. Characteristically Amy chose not to quit at this juncture. She had, after all, achieved the distinction of setting a new UK – India record and there would have been no shame in giving up. However, if she had done so she would now be a mere footnote in the annals of aviation history.

From Rangoon to Amy's next stop at Bangkok, is approximately 370 miles as the crow flies, an easy hop for the Gipsy Moth one might imagine but, with the monsoon weather now in full spate, not without its difficulties. After crossing the Gulf of Martaban in heavy rain-filled clouds and failing to locate the port of Moulmein, where she had been told she would find a pass through the mountains, she began to climb to 9000 feet to clear the peaks which lay ahead of her.

After crossing the Dawna range, flying blind for most of the time, she emerged on the far side to find she was well off course, even off her strip map. Instinctively, she did the most sensible thing: she turned due south and flew on, knowing that eventually she must reach the Gulf of Siam. It was not long before she recognised landmarks which led her to the Don Muang airport at Bangkok. The airport had been given international status in 1923; the first foreign airline to use it was KLM, who called in on their regular Amsterdam–Jakarta mail run. However, in 1930 it was still a grass airfield which became a morass of mud when the rains came. For Amy, exhausted, suffering from a splitting headache and with her limbs aching from the confines of a cramped cockpit, it became just one more hazard she had to negotiate.

As soon as she landed she was persuaded by one of the Siamese Air Force mechanics to go and lie down whilst her machine was pushed to the hanger for servicing. When she awoke it was already dark and she recalled:

> There were no lighting arrangements in the hangar, the only lighting was my own small torch. I had to give instructions in English to the Shell representative, who repeated them in French to a Siamese who translated them to the mechanics. I was pretty helpless and should have preferred to have cleared everyone out and got on with the job myself, but my headache was too bad . . . so I went back to the barracks. When I did return I saw two empty two-gallon XXL tins on the ground, and at once I asked where my supplies of oil were. 'Inside the engine,' I was assured. 'But how much inside?' I insisted. 'All that was sent for you,' came the

reply. This was four gallons and I knew the sump only held two. I asked whether the old oil had been drained out. No, it had not. I asked if there was any more XXL. No. Any similar oil? No. There was nothing for it but to have all the four gallons, plus old oil, drained out, filtered, and two gallons poured back.[5]

One can imagine the mental stress, apart from that of flying day after day, that Amy had to endure as a result of these difficulties of language and lack of skill, all of which could have spelt disaster. Much of the assistance she received when servicing her Gipsy engine was of a dubious quality, as witness David's help at Bandar Abbas.

When she took off from Bangkok on the following morning, and was settling back to fly the 900-mile hop to Singapore, she found that the top cowling over the engine had not been properly secured. It flew open, causing her to return to the airfield and make a landing in an aircraft heavily laden with a full load of fuel. At that moment she was literally sitting between two highly inflammable fuel tanks brim full, onto an airfield whose surface was dangerous to land on. Moreover, the delay involved in fixing the cowling now meant that she was unable to make Singapore before sunset.

The easiest route for her to follow would have been to fly down the west side of the Malayan peninsula, but she had been warned that rain had been falling in Penang for the last fortnight. This left her but one option, to make a 450-mile flight to Singora (Songkhla) on the east coast of Siam, halfway between Bangkok and her intended destination. Not long after leaving Don Muang for the second time, she ran into blinding rainstorms over the Isthmus of Kra, which caused her to hug the coastline, flying a few feet above the waves. The downpour became so heavy that she was forced to remove her goggles and squint over the side of the cockpit with rain stinging her eyes in order to keep sight of the coastline. At one point she found her compass needle swinging steadily around the four quarters of the dial, from south to west, to north, to east and back again to south. She discovered that she had been flying in a circle around a large flooded area!

As she neared Singora during the early afternoon, the weather cleared enough for her to see clean white beaches beneath *Jason's* silvered wings. Luckily, she landed on the safest part of the aerodrome, near to the sea where the sand was much firmer than in other parts. Even then, as she came to the end of her landing run, the machine almost nosed over as it entered very soft sand and she found it impossible to taxi through it.

Amy surveyed the scene and quickly decided that if she was ever to take off again the next morning, the biplane would need to be manhandled onto an adjoining road so that its firm surface could be used as a runway.

Fortunately the officials at Bangkok must have phoned Singora warning them of her arrival, for there was an English-speaking Siamese official waiting to meet her. This meant that there was little difficulty in mustering a group of about twenty men to lift the tail of the machine and push it to a spot alongside the roadway, where she could carry out a routine inspection.

She waited until nearer sunset for cooler air before attempting to look at the engine, by which time a crowd of inquisitive Siamese men, women and children had gathered around the roped-off area of where *Jason* now stood. They sat as if waiting to watch a performance, and began chattering, eating, drinking and licking ice cream. It must have reminded Amy of the time when, as a young girl, she sat on the beach at Bridlington to watch the Pierrots.

Struggling to undo fastenings which had been over-tightened by the Siamese mechanics at Don Muang the night before, Amy appealed through the medium of sign language and a spanner held aloft in her hand for a strong man to come and help her. The crowd were delighted when a young man stepped forward and undid the front cowling and the recalcitrant oil drain plug before resuming his place in the crowd. After this, whenever Amy wanted more help, she would call out, 'Where's my strong man?' and there would be peals of laughter as the crowd repeated 'Strong man. Strong man', and another volunteer would offer his services.[6] There is no doubt that this unexpected humour relieved some of the tension and tiredness that Amy must have been feeling.

Sir Francis Chichester, who made a similar flight to Amy's in a Gipsy Moth to Australia in the period December 1929 – January 1930, recorded in his book, *The Lonely Sea and the Sky*, that he was always grateful to have some assistance on the ground. The fatigue that he suffered on his flight was '. . . not because of the flying, but because of the unending negotiations and talk from the moment I landed to the moment I took off again, apart from the few hours spent in sleeping.' He too found that flying in monsoon weather was a frightening experience, where 'the nearer I came to the centre of the storm, the more the plane was tossed about . . . The water stung my forehead like hail, streamed into my eyes, down my chest and back.' These were the words of a very fit young man who took five weeks to reach Darwin!

To describe Amy's take-off on the following morning from the road beside the aerodrome at Singora as hazardous would be more than an understatement. Crowds had assembled from out of nowhere and were now lining the roadway to witness her departure. Although she tried to explain that they must stand well back, they remained where they were – within a few feet of her wingtips. As she pushed the throttle open she knew that if the aircraft swung on take-off, she would be responsible for the loss of more than a few lives, her own included.

Whilst learning to fly Amy had formed the habit of looking out over the left-hand side of the cockpit, for until the tail of the aircraft rose up there was no forward visibility. However, since having a rubber tube fitted to the front petrol tank air vent at Karachi, which now exited on the left-hand side of the machine, she had been forced to look over the right side. Now, as she hurtled down the road, ruddering the aircraft to keep it straight, she needed to look over both sides alternately to make sure she was nowhere near the crowds who were determined to have a grandstand view of the take-off. As she leaned her head to the left, completely forgetting the air vent modification, excess fuel spurted back and into her eyes blinding her for the moment. Never was she more thankful to be safely in the air. She gave a brief wave to the upturned faces beneath her and turned to head southwards.

After flying for four and a half hours under a band of low, grey stratocumulus, Amy could see two approaching aircraft. As they dipped their wings in salute and turned to fly alongside, she recognised them as Moth seaplanes which had been sent out from Singapore to meet her and to guide her down to the RAF aerodrome at Seletar.

As she touched down on the airfield and taxied towards the hangars, she could see a large crowd of expatriates waiting to greet her. Here were men and women dressed as if they were about to attend a summer garden party, the men wearing whites and their frilly-laced women carrying neat little parasols. It was all very British. They were eager to catch a glimpse of this slip of a girl from Yorkshire who was now hogging the headlines in their newspapers. As she stepped down off the lower wing, dressed in an oil-stained khaki shirt and shorts, a round of polite applause rippled through the crowd. She was already a celebrity.

Amy's family back home were delighted with their daughter's progress and had been in constant touch with her through the medium of cablegrams ever since she had reached Karachi. They were either sending messages of congratulation or, as in the case of her Insein mishap, messages of consolation and encouragement. Now she was at Singapore, they were mainly of deals that Will Johnson was making on her behalf with various newspapers for the rights to her story. To complicate matters further the Australian press was offering 'the lone girl flyer' lucrative sums if she would continue her flight from Darwin to Sydney so that she could make a tour of various Australian cities. In all these matters Amy relied on the judgment of her father, who in effect had become her business manager.

Amy left Singapore on the fifteenth day of her journey, resigned to the fact that she could no longer equal Hinkler's record. She was, however, still ambitious enough to want to put up a creditable performance, particularly in view of the Australian offers. Her initial idea was to take a short cut by flying a compass course across the Java Sea for 1000 miles direct to Sourabaya. It was an ambitious plan, which was soon replaced by a more cautious one when she flew into blinding rainstorms. Zigzagging around the towering canyons of black cu-nimbus to avoid the worst of the downpours, she was forced to abandon her compass setting and head for the coastline of the Dutch East Indies. She knew that it meant she would be unable to reach Sourabaya in one hop; indeed, it was unlikely she would even make Semerang that day.

Even Francis Chichester found that 'no two maps of the East Indies seemed to agree'. Small wonder that Amy failed to find the emergency landing field at Tjomal, near Pekalongan, on the northern coast of Java, approximately 240 miles from Sourabaya. After circling the area for some time she thought of landing on the beach, but settled for a piece of open, flat ground near a large building, which turned out to be a sugar factory. As she came in to land in poor visibility she struck some bamboo poles, which she later discovered were markers for the siting of a house to be built for the Dutch manager of the factory. Hitting the poles turned out to be a mixed blessing, since although they damaged the fabric on the lower wings, they did act as a brake in slowing the machine to a halt just short of a ditch, which could have upended *Jason* and caused irreparable damage to the propeller. Once more she realised that a hand of Providence had been upon her.

The next morning Amy was able to mend the slits in the fabric with some sticking plasters, whilst the Dutch staff at the factory arranged for the ditch to be filled in so that she had a longer take-off run. Motor fuel was brought from a local garage and filtered carefully into the tanks through a chamois cloth, giving her just enough range to make Semerang. She was always fearful lest particles of rust or dirt should clog a carburettor jet and cause a forced landing into what might be a highly dangerous area.

Before long she was flying over cultivated squares of flooded rice fields along the Java coastline, with the cloud-capped peaks of the Slamet range framed in the struts of her starboard wing. She landed at Semerang to refuel and was disturbed to note that the blades of her propeller were much the worse for wear. The continual battering they had received from monsoon showers had worn away the glued fabric surface of the blades, exposing them to indentation, much like the damage caused by the force of stone chippings.

One of the Dutch officials agreed to ring up a local man who owned a Moth to see if he had a spare airscrew. He had not, but he was prepared to have the one taken from his own machine and sent on to Sourabaya for her. Amy received this kind of help all along the route of her journey, and more so since her exploits began to be detailed daily in the newspapers.

At Semerang she was warned that the aerodrome at Darmo (Sourabaya) was quite difficult to find, and was advised to wait for the Dutch KLM flight to take off and follow it. Amy recalled:

> I agreed, and about an hour later took off after the Air Mail, a three-engined Fokker, whose pilot had promised to go as slowly as possible. At full throttle I could barely keep him in sight, and I did not dare lose him as he was not following my compass course, and treacherous volcanoes were, I knew, buried in the clouds unpleasantly near. I had to fly at full throttle, 2,200 revolutions per minute and a speed of 100 miles per hour the whole of the way. On throttling down to land at Sourabaya the engine spluttered and the propeller stopped dead immediately on landing.

She lost a whole day at Sourabaya for, having left the servicing and the fitting of the new propeller overnight to a Dutch mechanic, she found the next morning that when she ran the engine up on the chocks, it misfired when she switched onto the right-hand magneto. Upon dismantling the offending unit, she discovered that some of the teeth had sheared on the the flexible coupling which joined magneto to engine. The problem had been caused by oil dripping onto the fabric coupling, which in the intense heat had partially disintegrated.

No aircraft spares were available, but after much foraging around in Sourabaya, an ordinary car coupling was found and fitted. By the time the work was completed there was not the slightest chance of Amy even making a midway halt at Bima on the island of Sumbawa, 450 miles away, let alone reaching her intended stop at Atamboea in Timor. Maybe the delay was all for the best, for whilst at Semerang she had remarked, 'I want so very much to rest.' There is little doubt that at this stage of the journey she was living on her nerves, and the prospect of a second night's sleep at the home of her wealthy hostess came as a welcome relief.

A canopy of azure blue touched by faint wisps of delicate cirrus arched over Amy as she climbed into the cool air over Sourabaya early next morning. Ahead of her lay the 925-mile stage of her flight to the airfield at Atamboea.

Australia was now well within her grasp. As she headed out across the Madura Straits flying eastwards towards Timor, she looked down on the string of coral islands, many completely uninhabited, just a few of the 13,677 scattered across 3000 miles of ocean from the Andaman Islands off the Burmese coast to Darwin. She passed over the blue-green bays of Bali and then on to Lombok with its 5000 foot volcanic peaks dominating the skyline, before being reported over Bima on the island of Sumbawa at 11.30 a.m. Her estimated time of arrival at Atamboea was 4.30 p.m. and the radio station at Kupang was standing by ready to report a sighting.

Many people have asked how pilots cope with their bodily functions on a ten-hour flight such as this one. Stimulants such as caffeine, taken either in tablet form or in coffee, were often used to keep those pioneer pilots awake, but unfortunately they also acted as a diuretic. Drinking of any kind posed micturition problems within the confined space of a cockpit, especially for a woman. The usual method was to use a funnel connected to a rubber tube leading into a suitable receptacle – not the easiest of tasks, especially in turbulent weather.*

The afternoon light was fading fast as Amy flew across the 44-mile stretch of water between the islands of Flores and Timor. The monotony of the flight was only broken by the chore of hand-pumping the last of her fuel supply into the upper tank, which gave her approximately three hours' flying time, but no more. She longed for a respite from the incessant roar of the open exhaust and the noise from the buffeting of the wind around the open cockpit which she had endured for the past two weeks. They combined to bring on an extreme tiredness, to the point where she had difficulty in keeping awake.

She had been in the air for almost ten hours when suddenly she thought she could see the smudged outline of land on the horizon. Leaning forward she wiped the oil-smeared screen with her gloved hand to check if it was merely a bank of clouds. No, her eyes were not deceiving her, it was in fact Timor. As she flew along the northern coast she gained an exact fix on her position by recognising the distinctive shape of a small island directly opposite her destination. Turning inland she searched for the airfield but, as in her quest for the racecourse at Rangoon, she failed to find it. With her top tank fuel supply dangerously low and darkness now threatening, she became desperate as she skimmed the tree-tops looking for her elusive haven. She circled the area several times surveying the terrain before finally deciding to

---

* A Newark aviation mechanic once confided to a US aeronautical designer that the cabin of the machine of one celebrated aviatrix reeked of urine after she had made a long-distance flight.

make an emergency landing on what looked like the best of available open spaces. As at Tjomal, she had a miraculous escape. As she landed she narrowly missed mounds of earth, some as high as 6 feet, which she later discovered were huge anthills. *Jason* came to rest with one of the smaller mounds gently holding up its lower wing, but thankfully without damage.

Amy had landed alongside a cluster of mud and straw huts in the remote village of Haliloeli, from which sprang a surprising number of natives, some of whom at first glance appeared to be none too friendly. Amy drew her revolver, but was relieved when one man approached with a friendly salute. She could understand nothing he said except for the word 'pastor', and when he motioned her to accompany him she had the confidence to follow him as he led her to a small mission chapel. After a while a kindly pastor, who spoke only French, appeared on the scene and took her into the building, where he quickly rustled up a meal for her.

No sooner had Amy started eating than she was surprised by the sound of a car's horn outside the building, something she did not expect to hear in such a remote area. Moments later the Portuguese commandant of the airfield at Atamboea greeted her. He had heard her plane circling overhead and had followed her course out to the mission. It transpired that she had missed the so-called 'aerodrome' by only a few miles and had failed to recognise it because of recent bush fires which had blackened out the white cross which should have been marked on it. The lack of this identification mark, combined with the fact that the airfield did not have even a solitary hangar, was enough to make the search well nigh impossible for any tired pilot.

The radio station at Kupang had shut down at 11 p.m. with no news of Amy's safe arrival, and of course there were no telephone links between the village of Haliloeli and the outside world, so the London newspapers reported on the following morning that Amy Johnson was missing. For a few hours it must have been very worrying for her family, as frantic messages of cablegrams crossed between the UK, Java and Australia. Meanwhile, the Dutch authorities at Sourabaya were preparing to make an air search when the news eventually came through that she was safe. All of this only served to heighten public interest in her flight, not only at home but around the world.

The following morning there was a delay of some three hours whilst the pastor arranged for a few tins of motor spirit to be brought by donkey to the village, where it was eventually filtered into the Gipsy Moth's top tank. Whilst it was enough to enable Amy to make the few miles to Atamboea, it meant that there was now no chance of her reaching Australia that day, for her machine still needed servicing and a very careful inspection before she risked the Timor Sea crossing. The anthills were levelled by the local natives to give her every chance of clearing the trees at the end of the limited open space available, and she instructed them by sign language as to what they must do to ensure that she got off safely. After several attempts to instruct the men on how to restrain the machine from moving forward by holding onto the wings whilst she increased the engine speed, she finally got away.

The Timor Sea had been flown many times, but never before by a woman alone. Francis Chichester had made a similar flight from Atamboea to the one Amy proposed, when he aimed for Bathurst Island off the shore of the Northern Territory of Australia in order to reduce the water crossing to 320 miles. He had also given much careful thought to how he would ditch down into the sea so that he could release his inflatable dinghy if he suffered engine failure. His plan was to 'glide down as usual, and kick the plane across its path at the last moment to pancake sideways'. The sea was shark-infested, which caused aviators to think very carefully before they tackled it in a single-engined machine. The well-known long-distance flyer C. W. A. Scott, who had crossed it alone on two occasions, always carried an axe and vowed that if he was ever forced to ditch in it he would attempt to hack off a part of the wing and float with it, but if attacked by sharks he would not hesitate to use his revolver to take his own life. History proved that this was not just an idle threat.

Amy was about to set out on the most dangerous part of her flight without the precaution of any life-saving equipment, apart from a parachute which would have been of little use in a sea ditching anyway. However, she recalled that she made meticulous preparations before setting out.

> I did everything I could think of on the engine. I put in a new set of plugs I had kept specially for this stage of the journey, tightened up every single nut and bolt on the engine and machine, carefully examined all petrol and oil pipes and connections for the slightest hint of leakage, and did all the other jobs which come under that of a daily overhaul. I worked long after dark by the light of a big fire built for me on the aerodrome. I looked at my books to see that I had remembered everything. I was determined that I was not going to have any forced landing over the sea owing to my having left out any small thing.
>
> The next morning I set off on the last stage to Port Darwin. My engine ran beautifully, but halfway across it spluttered occasionally. It sounded like a slightly choked jet, and I wondered whether, in spite of all my precautions, some dirt or water had got into the petrol. Every time the engine spluttered I opened out quickly to full throttle to try to blow the

foreign matter out, and then quickly throttled down again. Black smoke escaped from the stub exhaust, but the engine never stopped.[7]

At the halfway point Amy sighted the Shell oil tanker *Phorus*, which had been forewarned to look out for the British aviatrix. She swooped low over its decks and was greeted by friendly waves from members of the crew as they recognised the G-AAAH markings on the plane's side. Checking her drift by the trail of smoke from the ship's funnel she adjusted her compass setting and made for Melville Island. At 3.57 p.m. on Empire Day, 24 May 1930, nineteen and a half days after she had set out from Croydon, the wheels of Amy Johnson's silver and green Moth touched down at Fannie Bay, just beyond the Darwin racecourse, and a legend was born.

Amy was given a tumultuous welcome from the public when she landed, for they quickly came to recognise the extraordinary achievement of this remarkable young woman. Admittedly, she had failed to beat Hinkler's record by four days, two of which, the delay at Sourabaya with magneto trouble and her failure to locate the blackened airfield in Timor, could have happened to any pilot, even the little Australian. Within minutes she was up on a hastily erected podium, looking tired but jubilant, to respond to calls of 'Amy, Amy', to which she replied, 'Don't call me Amy. Just call me Johnnie, that's what my English friends call me.'

A typical crowd scene wherever Amy landed in Australia – this one most probably at Longreach. (The Amy Johnson Estate)

News of her arrival went out along the submarine cable that ran from Darwin to Kupang and thence to London. Within hours messages of congratulation began to pour in from all over the world. One needs to remember that Britain at that time had a closely linked empire. King George V sent the following message to the Governor-General of Australia. 'The Queen and I are thankful and delighted to know of Miss Johnson's safe arrival in Australia and heartily congratulate her upon her wonderful and courageous achievement.'

Amy received personal telegrams from the King and Queen, Prime Minister Ramsay McDonald, the King and Queen of Belgium, Charles and Anne Lindbergh, Louis Bleriot and many other notables, as well as hundreds of letters from the public. Back in Britain the Post Office was delivering letters by the sackload to 85 Park Avenue, Hull; if people did not know her address, simply 'Miss Amy Johnson, England' was sufficient. The newspapers were describing her with eulogising epithets, which ranged from, 'Queen of the Air' to 'The Empire's Great Little Woman', whilst the songwriters were busy scoring songs such as 'The Lone Dove', 'Aeroplane Girl', 'Call Me Johnnie', and the most famous of them all, 'Amy', with its haunting refrain, 'Amy, Wonderful Amy' – a tune that began to be whistled by schoolboys and postmen alike.

The *Daily Mail* not only paid her £10,000* for the rights to her story, a sum that would equal £750,000 in today's currency, whilst the *Daily Sketch*, not to be outdone, launched a public appeal amongst its readers for donations of one shilling towards a fund to purchase a new aircraft for her. Added to these rewards were the advertising royalties that she would receive from manufacturers of the equipment she had used during her flight. Wherever she travelled in Australia (she had now decided to forgo the idea of a return flight to Croydon) and elsewhere, she was showered with personal gifts, many of which were of considerable value.

One is forced at this stage to ask why Amy Johnson became the international icon that she did – even the astronauts returning from their Apollo space trip to the moon in 1969 were not given the same popular acclaim. Although she was not a natural pilot – her frequent mishaps bore witness to that – her indisputable courage and her willingness to dirty her hands along with the

* This sum might seem insignificant when compared to the vast amounts paid to today's celebrities, but at the time it was the largest sum ever paid by a newspaper to a single individual for an achievement in aviation.

Ciss and Will Johnson with their daughters Betty and Mollie reading messages of congratulation at the news of Amy's safe arrival at Darwin. (Hull Daily Mail)

men in the hangar or the workshop greatly enhanced her appeal to the ordinary man and woman in the street. This was no spoilt, over-indulged, rich debutante or duchess who had learned to fly in order to seek some cheap publicity. Added to this her modesty of speech and simplicity of manner projected an image of 'the girl next door', and this became a major factor in her popularity.

Many people have asked exactly where Amy Johnson ranked in the annals of aviation history at the time of the completion of her Australia flight. There had been many notable women pilots long before her, such as the French-born Raymonde de Laroche who, in 1910, became the first woman to be issued with a pilot's licence and the renowned American aviatrix, Harriet Quimby, a New York journalist who, on 16 April 1912, became the first woman to fly alone across the English Channel. Others, such as Lady Mary

Bailey and Lady Mary Heath, had made their mark, but none could compare with the diminutive Yorkshire woman for the endurance and bravery she had displayed on her solo flight to Australia.

Throughout her flying career, Amy's main rival was to remain the celebrated aviatrix, Amelia Earhart. However, although the American held altitude and trans-continental flying records within the USA, she had at the time of Amy's Australia flight only achieved international distinction by virtue of being the first woman passenger to be flown across the Atlantic. Much more was to follow later in the careers of both women.

'Call me Johnnie'. (Harry Carlton)

Chapter Ten

ON THE CAROUSEL

xcitement mounted in the Johnson household as camera crews began to arrive from Gaumont Movietone to interview the family and film the home of their famous daughter. Within a few days Will, Ciss, Mollie and Betty were sitting in the best seats at their local cinema, watching themselves on the screen, along with shots of a jubilant Amy being fêted by enthusiastic crowds in Australia. These were heady days for the whole family, but the excitement did not deflect Amy's father from bringing his considerable skills as a businessman to bear in negotiating the best possible terms with the London newspapers on behalf of his absent daughter.

Ever since Amy had arrived in Karachi there had been a constant flow of cablegrams from her father advising her on all the aspects of her flight. At Calcutta he was suggesting that she should spurn the enticing financial offers she was receiving to make a tour of major cities in Australia, fearing that a long absence would cause interest to wane in Britain, and urging her to return home as soon as possible after reaching Darwin. He did not, however, push her to continue the flight after her mishap at Rangoon, for there were constantly phrases in the cablegrams such as 'Don't overdo yourself', and 'Be careful.' At Sourabaya, just before the ordeal of the Timor Sea crossing, he cabled, 'Please don't attempt sea-passage unless feeling quite fit and your engine dependable. No hurry, be careful.' It seems that even before she reached Darwin, he would have settled for her returning home anywhere along the route, once she had beaten Hinkler's UK–India record, but his daughter was not prepared to give up that easily.

Amy seemed more than happy to have her father act as her business manager until she returned home. When she arrived in Darwin he was already advising that he had negotiated with the *Daily Mail* for her to return to England by 1 September, in order for her to take up a position on their staff as an official lecturer and air correspondent until the following April. It is evident that at this stage neither he nor Amy realised the tremendous strain that a UK tour the paper was proposing would place upon her, especially since she had now decided to go ahead with the Australian offer. In addition to the offer of £10,000, the *Daily Mail* were now prepared to give her a further contract, whereby she would receive 50 per cent of fees received for her appearance in movie and talking pictures. Offers had also been made for Amy to appear in various theatres around Britain, but these were wisely spurned by Will Johnson, who cabled her to say that he had refused them in order 'to avoid making you cheap or putting you on show'. He also warned her to be most careful when dealing with the Australian press, so that she did not contravene the *Daily Mail* agreement. This meant that she should not give detailed interviews or make lengthy speeches.

The Australian and British newspapers were not the only ones seeking a slice of the action from this money-spinning heroine, for both of the major oil companies, Castrol and Shell, were on the scene as soon as Amy landed at Darwin. Cyril Westcott was the General Manager of Wakefield Oils in Australia and he had arranged for his northern representative, Captain S. W. Bird, to be flown up from Brisbane to meet and chaperon Amy throughout her Australian tour.

The Qantas pilot allocated for this flight was tall, athletic Londoner, C. W. A. Scott, who was soon to achieve world-wide fame as a long-distance aviator. He had begun his flying career with the RAF at Duxford, where he had qualified as a fighter pilot in 1923, and in the following year had become the RAF's heavyweight boxing champion.

Charles William Anderson Scott was a strange mixture of a man. There was a gentle side to his personality – the poet, the classical pianist (his father had been a Master of the King's Music) – and yet there was also a wild and arrogant element. In Longreach on a Saturday evening some of the rougher types from the sheep stations would come to town, hear about the musical poet and, being ignorant of his boxing prowess, seek to pick a fight with him. One observer from those days recalled, 'Scott might have been good at Beethoven but he could also knock over an eighteen stone shearer without even putting his beer glass down.'

Shell Oil were not allowing Castrol to monopolise Amy, having already given instructions to the Captain of the *Phorus* to act as a navigational marker for the aviatrix as she flew across the Timor Sea. They had also sent their representative, Flight Lieutenant Harold Owen of the Royal Australian Air Force, up to Darwin in the company's Gipsy Moth to act as an escort during her flight across the northern territories. Scott and Bird, as representatives of Lord Wakefield, were not particularly pleased that one of their rivals should accompany them on their flight to Brisbane, but there was little they could do about it.

Scott appears to have adopted a churlish attitude towards Amy right from their first meeting. Maybe it was because he resented being merely an escort to the unknown slip of a girl who was receiving so much adulation. Behind her back he was harshly critical of the state of her engine, saying that it had been poorly maintained and was in an appalling condition. He did not seem to appreciate the circumstances under which she had worked at the end of each day's flying. It had meant having to work while dog-tired under poor lighting, and while fending off pestilent insects. Owen was the most sympathetic and understanding of the three men, and he appears to have carried out the major part of the maintenance of Amy's machine whilst she was at Darwin.

The three planes took off from the tussocky grass field of the Fannie Bay racecourse on the morning of Monday, 26 May, just two brief days after Amy had arrived in Australia. Leading the way, and setting an unnecessarily scorching pace, were Charles Scott and Captain Bird in the speedier Qantas four-seat cabin-biplane, followed by Amy in *Jason* and Owen in a similar aircraft. Scott was well qualified to navigate across the featureless landscape which lay beneath them, for he had long been familiar with the route as an airline pilot, but on this occasion he seemed to make things unnecessarily stressful for both Amy and Owen, since they both had difficulty keeping him in sight. His excuse for the fast pace was that he was obliged to keep to the tight time schedule that was laid down for him.

They landed at Daly Waters to refuel and were quickly back in the air again for their first overnight stop at Alexandria Station. As the flight progressed over the four-day period, the number of people waiting to catch a glimpse of 'Australia's Sweetheart' increased daily. Stops were made at Conclurry to refuel, and Longreach for another overnight. Such was the wild enthusiasm of the crowds that Amy was on occasions man-handled from the machine and carried to a makeshift dais, where she was expected to give a short speech about her flight from England. Gifts such as medallions, brooches, clocks, watches, cups, boomerangs and trays were presented to her. No doubt these would have aggravated her escort pilot even more, for they would need to be transported in the cabin of the Qantas machine.

By the third day out from Darwin, the stress of the journey was beginning to tell on Amy. Not only was she extremely fatigued, but her period had started. It would have needed little imagination on the part of her escorts to deduce that a woman would have timed her three-week flight so as to avoid her monthly cycle, and that she would now be into the fourth week, but if they did realise that this was her problem, they showed scant regard for it. In any event, somewhere between Longreach and Charleville, Amy lost sight of her two escorts and decided to deviate from her intended course and follow the railway line down to Quilpie, some 120 miles to the west of Charleville, where she was able to rest for a number of hours before proceeding. The entry in her flying log for 28 May simply reads, 'Quilpie–Charleville, 1 hr 40 mins. Landed after dark.'

Disaster followed on the final stage of the flight to the Eagle Farm airfield at Brisbane. Amy had already made a refuelling stop at Toowoomba, where again she had been overwhelmed by the crowds waiting to greet her, and by the time she took off for her final destination, she was really in no fit state to be flying. Her Wakefield and Shell escorts were already waiting for her when she arrived over Brisbane, where a crowd of 20,000 were assembled to greet her. To those on the ground it appeared that she made her landing approach with far too much height, and then attempted to correct it by making a steep side-slip. As a result, by the time her plane touched down she was heading for the boundary fence, and much to the dismay of all the onlookers her aircraft struck the fence and somersaulted over onto its back and into a field of maize.

Jason over the Brisbane River and heading towards the airfield at Eagle Farm. A few moments after this picture was taken she made her near disastrous landing, one which seemed to endear her to the people of Australia more than ever.
(The Amy Johnson Estate)

The crumpled wreckage of Jason *after Amy overshot the airfield. The Gipsy Moth was shipped back to Britain where it was repaired and made ready for her homecoming at Croydon on 4 August 1930.* (The Amy Johnson Estate)

The incredulous crowd watched as Amy unstrapped herself and slipped out from beneath the crumpled wreckage of her aeroplane. There had been many official arrivals at the airfield before, but none quite like this one. To have failed in front of such a large number of people, and particularly under the critical eyes of Scott and the two oil representatives, must have damaged her pride to the point of humiliation, but with great self-control she managed to put a brave face on the whole incident.

Clad in riding breeches and calf-length boots, topped by a sleeved pullover, shirt and tie, she climbed the steps of the dais to make a speech which immediately endeared her to the people of the city. Here was a woman who never admitted failure and the crowd were quick to recognise and applaud this characteristic. It subsequently became the hallmark of the Amy Johnson legend, showing that she was no quitter. She earned the reputation of someone who had the ability to get up from a near disastrous fall and go on.

One must remember that Amy had very little experience in public speaking, apart from some minor speeches she had made as a student at university. She now had to think quickly on her feet, for there was no time for careful preparation. To add to the stress, there were numerous interviews with reporters, not all of whom were on her side. It was the time of the Great Depression, when unemployment was a crippling factor for many people, and Australia was no exception. Soup kitchens there might be, but social security as we know it today was still a long way away. An astute paper, looking for a good story, could easily dismiss the adulation and home in on the fact that this young woman had earned more in three weeks than many Australians would earn in a lifetime.

One such news magazine was *Smith's Weekly*, a popular Sydney 'rag' which lampooned Amy as the 'Gimme Gimme Girl', largely as a result of Cyril Westcott demanding a fee of £100 on Amy's behalf for personal appearances. Whether the demand was initiated by Amy is hard to say, although one can detect a touch of cupidity in her responses to requests for appearances*. She may well have made the most of her opportunities, but who could blame her, for she had known relative poverty during her time in London before she became famous – at one time her only pair of shoes were worn through so badly that when she arrived for work at Peter Jones she was forced to stand around all day with wet feet. Although she managed to put the criticism of the Australian article behind her it did upset her. It became just one more stress factor she was called upon to endure.

The day after her arrival in Brisbane Amy was paraded through the streets of the city in the back of an open-top car, waving and smiling to the cheering crowds. Leading the procession were mounted police, whilst alongside Amy sat Cyril Westcott and a poker-faced, bespectacled Captain Bird, both looking as if they would rather be somewhere else. On Amy's part it was all a façade; she was now merely playing a role which was about to overwhelm her. As soon as she gained the privacy of Government House, where she was being hosted, she broke down in tears.

It was very much a case of Wakefield Oils being the piper who had to be paid, and Amy had to dance to their tune. Sadly she had been the willing victim. News of the strain she was now under must have percolated through to her parents for on 3 June, the day she was awarded a CBE in the King's Birthday Honours List, she received a cable from her father which read, 'Alarming newspaper reports concerning your health. Cable us exact condition.'

* Whilst on her honeymoon in Scotland she was asked to appear on the stage during a summer show at the Barrfields Pavilion in Largs, but she declined unless she was paid a fee.

When opportunity knocked, Charles Ulm's hand was always somewhere near the door, ready to swing it open. As soon as he heard that on doctor's orders she was to continue the rest of her journey to Sydney as a passenger rather than pilot her own machine, he immediately contacted Wakefield's Cyril Westcott. He seized on the prospect of free publicity for Australian National Airways by offering to place one of the company's airliners completely at her disposal. By the time Major Hereward de Havilland had flown his company's DH Hawk Moth up from Melbourne with spares for the damaged *Jason* and offered to fly Amy down to Sydney, Westcott had already accepted Ulm's offer. Amy did, however, accept the Major's offer to fly her on to Canberra, Melbourne, Adelaide and Perth.

The calm blue eyes of co-pilot James Allan Mollison glanced from the flickering dials of the instrument panel of the Avro tri-motor *Southern Sun* into the haze of the Australian landscape as he took over the controls. Vacating his seat alongside him, first pilot Charles Ulm, Director of the newly formed Australian National Airways, moved to the rear cabin to stretch his legs and to chat with some of his eight passengers. Ulm was anxious to encourage his customers to become regular users of his airline, rather than to fly with his rivals, Qantas and Western Australian Airlines. Conversation was not easy, for in 1930 sound-proofing on aircraft was a distant luxury for passengers and crew members alike, and Ulm had to shout to be heard above the noise and roar of the three Armstrong Siddeley Lynx engines. Most of his passengers that morning were sheep farmers, businessmen and their wives, travelling from Sydney to Brisbane on one of ANA's four-times-a-week service between the two cities.

The airline had been founded in 1929 by Charles Kingsford Smith and his fellow Australian Charles Ulm with a share capital of £85,000, of which more than half had been sunk into the purchase of five tri-motor monoplanes. These British-built Avro 10s (so called because they carried a crew of two and eight passengers) were named *Southern Cloud*, *Southern Moon*, *Southern Sky*, *Southern Star* and *Southern Sun*. Together with an existing Dutch-built Fokker, *Southern Cross*, they comprised the company's fleet of aircraft. ANA's primary objective had been to establish and operate a regular mail, passenger and freight service between major cities, including Melbourne, Canberra, Sydney and Brisbane.

Kingsford Smith and Ulm had already achieved lasting fame by becoming the first to make an aerial crossing of the Pacific from the USA to Australia.

With navigator Harry Lyon and wireless operator James Warner, they had set out on 31 May 1928 from Oakland, California, to fly their *Southern Cross* over 7389 miles of inhospitable ocean to Brisbane, where they arrived nine days later to be acclaimed as international heroes. Now, having decided to go into the airline business, they had recruited some of the best pilots available in Australia to fly their new ANA machines. Amongst them were Travers Shortridge, Gordon Taylor, Jerry Pentland, P. W. Lynche-Bloss, George 'Scotty' Allan and his fellow Scot, Jim Mollison.

The company's inaugural flight was on 1 January 1930 and for the first three months Jim Mollison, very much the junior pilot at the time, was allocated to fly with senior pilots Allan and Shortridge on the airmail run from Sydney to Brisbane. It seemed that the company needed to prove their ability to fly the mail without mishap before they managed to encourage fare-paying passengers. During these proving flights, Mollison recorded the monotony was sometimes broken by the sight of the dark sinister shapes of sharks just below the surface of the water. Often they would see a shoal of five or six of these man-eaters, and if they happened to be near the shore they had an arrangement with the bathing beaches whereby they would fire off flares as a warning.

It was not long before Jim Mollison gained a reputation amongst his colleagues not only as an excellent pilot, but also as having a keen eye for the ladies. Some put it less politely by describing him as 'skirt crazy'. Although he was born and bred in Glasgow, he had what was described by a wartime friend as an impeccable Oxford accent, which he acquired during his public school education and a five-year short-service commission in the RAF. He was known for his dry sense of humour and had a reputation for having a very good chat-up line with any attractive girl that he chanced to meet.

Former ANA ground engineer, Geoff Wells, recalled that it was

James Allan Mollison. (J. Capstack)

not unusual for him to be approached in the Essendon, Melbourne, office, by an attractive young female who asked him to tell Captain Mollison as soon as he landed that she would be waiting for him in her car at the end of the road. When he did pass the message on, the reply would often be, 'Oh! gosh, I'm going out this way. If she comes back, tell her I've gone. Tell her anything.'[8] On one occasion there were actually two girls waiting at the same time one in a red Oldsmobile at the end of the road, and another in a Studebaker at the other end.

On another occasion Mollison eyed up a rather stunning-looking passenger who had boarded his aircraft, accompanied by a tall, powerfully built man. As he went up and sat down in the front cockpit, he quietly confided to 'Scotty' Allan that this girl was the loveliest creature he had ever laid eyes on, and that he intended to fix a date with her as soon as they landed at Melbourne. Allan declined to comment at first, but when pressed for an opinion he said that he felt it would be most unwise in view of the fact that he knew that the man with the girl was her husband. Mollison did not see that this posed any real problem until Allan informed him with a wry smile that the man was called Lewis – 'Strangler' Lewis, Australia's current champion wrestler!

Although Mollison's enthusiasm for dating women passengers was dampened on that occasion, it did little to discourage him just a few weeks later.

There is no doubt that Mollison liked what he saw when he first met Amy as a passenger on the *Southern Sun* at Brisbane's Lytton airfield on 4 June 1930. He considered himself a connoisseur of pretty women, and as she stepped on board, dressed in a neat green suit, she must have created a unique impression upon him, for here was a woman who combined sexuality with unquestionable bravery. Physical courage was always something he admired in women.

We know little of her initial reaction to him – the man she was eventually to marry – when they first met. Hans Arregger was merely a distant memory and maybe by now her bitterness towards men had dissolved in the intoxicating and heady world into which she had been thrust. She knew how to flash her luminous blue-grey eyes and play the *allumeuse*, for she was not unaware that men found her attractive.

During the five-hour flight Amy sat for most of the time chatting with Charles Ulm and his wife Josephine, who was acting as the air hostess. With

Captain Bird they sat around a small card table decorated with poinsettias, whilst Mollison flew the airliner. Amy made no mention of the co-pilot in her log-book entry, merely recording that Ulm was the official pilot and that she took over the controls for part of the trip. She had never flown anything larger than her own single-engined biplane before and no doubt as she sat alongside the young Scot, he turned on the charm as he explained things to her. We know for sure that he managed to get Amy to promise to dance with him that evening at the Wentworth ballroom in Sydney. She was to be the guest of honour at a reception to be given by the Australian Flying Corps Association and he obviously knew enough contacts in that circle to wangle an invitation for himself.

Jim Mollison arrived, suitably attired in evening dress, to claim the promised two dances. There was just one snag: she was being closely shadowed by her chaperon, who happened to be the State Governor, Air Vice-Marshal Sir Philip Game. As Mollison made his way across the floor towards Amy he was intercepted by her imperious guardian, who announced, 'Miss Johnson is extremely tired and will not be dancing any more this evening.' Mollison was quick to realise that he was outgunned. Junior ex-RAF officers do not take on AVMs and win. Not wishing to create a scene, the crestfallen Scot retired gracefully, and no doubt sought refuge in more than a few drinks at the bar. He thought it unlikely that he would ever meet Amy Johnson again, for he knew that quite soon she was to move on and within a month would leave for England.

The frenzy with which Amy was greeted in Sydney was most probably the highlight of her six and a half week stay in Australia. Will Johnson's diary entry for the day she arrived in the city simply read, 'Amy having great fun in Sydney.' Even as the *Southern Sun* approached the aerodrome at Mascot it was met in the air by several light aircraft which were being flown by local women pilots to welcome their heroine from the mother country. Not only was Australia far more air-minded than Britain at that time, but it was also the home of an incipient feminist movement, which saw in Amy a potential propagandist for their views. For them she became a symbol of women's emancipation.

Amy was driven at a snail's pace through the densely crowded city, waving, laughing and smiling, to Sydney's Town Hall. Press photographs show the reception hall packed with women standing in the galleries and perched wherever they could catch a glimpse of their new idol. In the lower part of the hall, cloche-hatted women in low-waisted dresses were seated at long dinner tables to listen to the speeches of welcome and reply. There can be little doubt that the women as well as the men in all the major cities took this young woman to their hearts. They warmed to her personality as well as to her achievements in aviation.

Her success in winning the hearts of Australians, in spite of some of the criticisms of certain sections of the press, was repeated in Melbourne, Adelaide and Perth. Much like royalty she was obliged to perform, to smile and to make impromptu speeches, even when she did not feel up to it. Unlike royalty however, who have life-long training for their role, the girl from Hull had to rely purely on her intuition and common sense. She admitted that when she first arrived in Darwin and was hauled up on to the platform, she felt she was going to faint with fear. Her shyness was taken by some as a form of snobbishness and she had to work hard to overcome her natural reticence in public. The only complaint she was known to voice was that she did not meet enough people of her own age, for the dignitaries she was hosted by were largely from the older generation.

Crowds thronging Swanston Street, Melbourne, to greet the aviatrix as she made her way to an official welcome at the Town Hall. (The Amy Johnson Estate)

Inevitably, there were times when Amy felt unable to keep up with the heavy schedule that had been arranged for her; constant attendance at official functions, dinners, galas and society meetings left her emotionally and physically drained. She began to realise that she was sitting on a carousel where the music refused to stop. It was as if she was being fed to the public very much like a commodity, and yet there was little she could do about it, having freely entered into contractual agreements with Wakefield Oils and the press.

There was evidence that she was dealt with quite roughly on at least one occasion when she had refused to make an exhibition flight in a German aircraft at a Perth Aero Club, thinking that it would be a poor advertisement for British aviation. Ian Grabowsky, the Sales Manager for the de Havilland Aircraft Co. in Australia, happened to be passing Amy's hotel room when he overheard her crying, and the Wakefield representative, Bill Brasch, shouting, 'You're our servant – we bought you body and soul and you'll bloody well do what you're told and what we arrange.'[9] Grabowsky went to the rescue and entered the room without knocking to find Amy lying on the bed sobbing and Brasch standing over her in a threatening posture.

The only ripple that raised a few public eyebrows occurred when an over-amorous admirer stepped out from across a street in Perth and attempted to kiss Amy. Without much hesitation she lashed out and gave the man a hefty slap across the face, causing him to scurry back into the crowd. It happened a few days before she was due to sail on the SS *Naldera*, and no doubt by this time she had received more than enough of the public's attention. One admirer put a rather humorous slant on the whole episode by presenting her with a pair of boxing gloves just before she left for home. It was a gesture which would have appealed to Amy, for a recent newspaper report headed 'Boxed with Apprentices' had quoted Jack Humphreys as saying, 'She's a strong girl, but we made her tough. She often used to take a turn with the gloves and get knocked about all over the hangar.' Quite obviously the audacious street Romeo who attempted to steal a kiss had not read the article.

A press photograph taken just before she boarded the P & O liner on 7 July and most probably the last to be taken of her Australian tour, shows a porter on the Fremantle dockside wheeling a large wooden crate containing the many gifts she had received. The crate was marked, 'Miss A. Johnson, c/o "Daily Mail", London, England.' It was just the kind of photograph that the editor of *Smith's Weekly* might have chuckled over.

Rain beat a steady tattoo on the windscreen of the the Imperial Airways Argosy airliner *City of Glasgow* as it made its landing approach over the suburbs of London on 4 August 1930. The leather-coated pilot, Captain Jimmy Youell, sat hunched over the controls in the open cockpit as it touched down on the undulating grass surface of London's Airport at Croydon. It was a Bank Holiday Monday with weather to match, and although it was only 9 o'clock in the evening it was almost dark. The poor weather and the fact that the airliner was several hours overdue had not deterred the vast crowd of 200,000 now waiting to greet Amy upon her return.

The official reception committee, which included the Air Minister Lord Thomson, Lord Wakefield, Sir Sefton Brancker, Miss Margaret Bondfield (the first woman to hold office as a minister), the Mayor and Sheriff of Hull and Amy's proud parents, was already positioned on the welcoming platform as Amy stepped down from the plane. Alongside the dais stood her beloved *Jason*, which had been shipped back earlier, after its refurbishment by de Havilland in Melbourne.

To the accompaniment of boisterous applause, Amy made a short speech, full of patriotic fervour. It struck an instant chord with the public when she said, 'I want to show by my flying how much I love England and its people, how glad I am to be back home, how proud I am to be a member of our own great Empire and how deep is my gratitude to you all.'

That night she took London by storm. Crowds, many of whom had waited for three hours in the rain, lined the streets four and five deep along the 12-mile route that took her convoy of cars from Croydon via Streatham, Brixton, Westminster Bridge, Whitehall, the Mall and Constitution Hill to Park Lane's Grosvenor House Hotel, where she and her family were to be guests. Some have estimated that the crowd was as many as a million.

It was a welcome which vied with that given to royalty, for even before her homecoming the young woman from Hull had stirred the imagination of the people of Britain. She was one of them. Although her name had become synonymous with courage and determination, it was the fact that she was so unpretentious and ordinary which seemed to appeal most of all to the public. People felt safe in greeting her familiarly as 'Amy', in an age when informality of address was not the norm and when one never used a person's first name unless one was very well acquainted with them. It was something almost unique. One is tempted to ask what would have happened if it had been Lady Mary Heath who had made the first solo flight to Australia by a woman. It is doubtful whether the titled lady would have received the same kind of response.

The champagne must have flowed freely that evening in the Grosvenor, for Will Johnson's diary records that the family did not get to bed until the early hours of the morning. The following afternoon Amy and her mother were out shopping in Oxford Street for clothes suitable for the numerous official receptions that were lined up for her. However, even before the first of these events, the Savoy Hotel luncheon on the following day, she was almost on the edge of a nervous breakdown. A doctor and a specialist were called and with the help of some form of medication she allowed herself to go on with the heavy schedule that lay before her. She now found herself on a treadmill and had little option but to go on.

Matters were made worse by the fact that her *Daily Mail* contract, which had been negotiated in all good faith by her father, required her to make a much more strenuous tour of the country in her aeroplane than she had ever imagined. The newspaper had appointed their air correspondent, William Courtenay, to take charge of the arrangements, and he had gone about his duties in an over-zealous way, without appreciating all that Amy had endured over the past six months. Her nerves were shredded and the thought of making a tour of some thirty-nine towns throughout the country was more than she could bear.

In addition to arranging the air tour, Bill Courtenay became what he described as Amy's 'man Friday', which meant that he handled most of her private affairs, including opening and reading the 500 letters a day that she was receiving as fan mail. Amy had many ardent admirers, some of whom wrote to her with romantic intentions – curiously enough, even in those so-called innocent days not all of them from men. Acting very much as her moral guardian, he recalled: 'Then there were strange letters from equally strange and perverted females who wished to make clandestine appointments through advertisements. For what purpose they wanted the girl may be guessed. Naturally I consigned those letters to the flames.'[10]

There was a glittering array of celebrities from the world of aviation, politics, art, science and sport to honour Amy at the Savoy Hotel luncheon on 6 August, when the Hon. Esmond Harmsworth, son of Lord Rothermere, presented her with the *Daily Mail*'s cheque for £10,000, and a gold cup from the Youth of Great Britain. Amongst those present were Sir A. V. Roe, Air Vice-Marshal Sir Sefton Brancker, Sir Philip Sassoon (representing the Royal Aero Club), Squadron Leader Bert Hinkler, M. Louis Bleriot, Mrs Victor Bruce and Miss Winifred Spooner. Others included were Noël Coward, Ivor

Novello, Val Gielgud, Malcolm Sargent, John Barbirolli, Alfred Hitchcock, Cecil Beaton, J.B. Priestley, Beverley Nichols, Evelyn Waugh, Gordon Selfridge and the Hon. Max Aitken. It says much for Amy's resilience that she was able to face and win over such an awesome group of people with so little experience.

Amy with Charlie Chaplin, Lady Nancy Astor (the first woman to become a Member of Parliament) and George Bernard Shaw. (The Amy Johnson Estate)

Amy knew that if she was to fulfil her *Daily Mail* contract, her most pressing concern was to regain her nerve in the air, for ever since the Brisbane crash she had not piloted an aircraft alone. In order to do this she sought the assistance of her erstwhile flying instructor, Captain Valentine Baker, who was now employed at the Heston Air Park. Maybe it would have hurt her pride too much to have gone back to Travers at Stag Lane and admitted that she needed someone to restore her confidence. Her flying log records that on the Friday and Saturday of her first week back in the country, she made three half-hour flights in one of the Heston club's Moths, after dual with Baker.

The first of her flying engagements began when she flew *Jason* to Hull on the afternoon of Monday, 11 August. She landed at the city's municipal airfield at Hedon, which was thrown open to the public for the occasion. The welcome in her home town was every bit as frenzied as the one she had received in the capital. She was driven to the city centre through streets lined with schoolchildren cheering and waving. One witness recalled, 'My wife was outside the City Hall amongst the crowd awaiting her arrival. They knew she was coming in by air, but so little did they know about aviation, most of the crowd imagined that she would somehow drop out of the sky in front of City Hall!'[11]

From the balcony of City Hall Amy told the crowd that she had been given a purseful of gold sovereigns by the people of Sydney, and that she intended to use them to purchase a gold challenge cup to be presented annually to a child of Hull for acts of bravery or courage. Her promise was fulfilled when in 1932 a trophy was inscribed 'The Amy Johnson Cup for Courage'; it has been awarded annually ever since. After her speech her entourage, led by the Lord Mayor, moved on to the Guildhall where she was given an official civic reception.

Amy's stay in her home town lasted barely four days, but it was long enough for her parents to notice warning signals that their daughter was in much need of rest. They managed to convince her to take a holiday with her sister Mollie immediately after a presentation which the *Daily Sketch* and the *Sunday Graphic* would be making in Hyde Park the following Saturday. The two newspapers were determined not to be outshone by their rival the *Daily Mail*, and they had been running a campaign ever since she landed in Australia for readers to donate money to buy her a new aeroplane in place of her second-hand one. An article in the *Daily Sketch* at the time was headed 'Women's Shillings for Amy's Plane'. It seemed that a whole spectrum of people, including peeresses, typists, actresses and famous airmen, were sending in their donations.

Whilst it could not be said that Amy's life had gone from rags to riches, it had most certainly gone from relative poverty to overnight wealth. It is difficult to assess just how much she was worth by today's standards but if one includes the cheques from the *Daily Mail*, the gift of a brand new Puss Moth from de Havilland, the Moth from the readers of the two newspapers, the brand new MG 18/80 Mk 1 black saloonette from William Morris (later Lord Nuffield) and many other gifts too numerous to mention, it must have come to about £16,000. Allowing for the £300 that she paid back to her father for his share of *Jason*, plus the money with which he had subsidised her flying tuition, the total may not strike one as particularly impressive. However, working on an average wage at that time of £208 per annum and comparing it with the current average of £20,916, it made her worth at least £1.6 million at today's values.

Over 100,000 people assembled in London's Hyde Park in glorious sunshine on the Saturday afternoon. She was escorted through the waiting

ABOVE:
Amy with the M.G. Saloonette (18/8, Mark 1) which was a gift from Lord Nuffield.
(The Amy Johnson Estate)

Amy alongside Sir Sefton Brancker as she greets her fans in Hyde Park just before she was presented with a brand new DH Gipsy Moth (G-ABDV) by the readers of the Daily Sketch *and* Sunday Graphic.
(The Amy Johnson Estate)

crowds by a dapper and monocled Sir Sefton Brancker, smiling and waving to her well-wishers. Ironically, the last time she had been in this same park was when she was living in digs nearby in 1927. On that occasion she had been watching the riders on what she described as their 'ripping' horses cantering along Rotten Row, and she had wondered to herself if their owners, with all their wealth, were really as happy as they looked. Little did she ever expect to be in a position to find out. As she mounted the platform to receive the gift of the *Daily Sketch* aeroplane, a DH Moth which stood resplendent in its pristine livery beside the dais, a thirty-piece brass band alongside her began to play 'Amy, Wonderful Amy'. The crowd loved every minute of it and they warmed to her as she responded to their generosity and welcome.

Amongst the vast crowd gathered on that day was Hans Arregger, with his mother and one of his sisters. It was the last time Amy was ever to see Hans and it appears that little more than polite conversation passed between them. One would dearly love to know something of their thoughts passing through each of their minds as they met. Was he thinking, 'Am I responsible for all this?' or 'What have I missed – how wrong can one be?' Was she thinking, 'Is this really the man on whom I was once so dependent?' They were now living in two different worlds, where perhaps neither were ever destined to know true happiness.

Mollie came down from Hull by train to meet Amy and the two sisters set off after the presentation for a week's motoring holiday in Wales, staying at various small hotels in the hope that they would not be recognised. They travelled under assumed names as Mollie and Ann Jones, with Amy playing the part of bluestocking with horn-rimmed glasses and a severe hairstyle to match. The ploy hardly worked, for press photographs of Amy being presented with the MG clearly showed its MG 720 registration plate – it was a complete give away, and they were soon spotted and pestered by reporters. Privacy was now a thing of the past, and Amy had to learn to live with it. In a letter to her friends Sheila and Mabel Glass from the St David's Hotel in Harlech on 22 August, Amy complained, 'I don't like all the publicity and only want to fly and travel.' In another letter, written eight days later, she told them, 'From what I can see they [the press] talk some awful rot and grossly exaggerate.'

There is further evidence that Amy was in no fit state to begin the air tour that the *Daily Mail* was now proposing. She was due to appear at

Eastbourne on the Tuesday afternoon, just two days after she finished her holiday in Wales, and Bill Courtenay had arranged for her to land at a small flying field near to the town's Hampden Park. The Brisbane crash was still very much in her mind, so much so that she decided to rehearse her landing at the southern coastal resort on the Monday evening. Without an audience in attendance she made a successful landing, but on the following day, when thousands were present and she was to be officially welcomed, it was a different story.

In order to put on something vaguely resembling an air show, it had been arranged that Amy would be accompanied by three other aircraft flown by local pilots. The four planes arrived over the selected site at the due time and the three local pilots broke away to perform some aerobatics a safe distance from the airfield. However, much to the dismay of Bill Courtenay and the more knowledgeable people present, when Amy came in to make a landing she made a complete hash of things. On the first three attempts she misjudged and touched down far too near to the crowd, causing her to open the throttle and go round again. She overshot the field much as she had done at Brisbane. The majority of onlookers no doubt imagined that she was just making low-level passes and giving them value for their money as she roared over their heads. It was only on the fourth attempt that she got it right.

Again, as with her mishap at Brisbane, she kept her cool and put on a brave front. Pictures taken at the time show a cluster of uniformed bobbies running alongside her aircraft as she taxied across the field towards a welcoming committee of local dignitaries. Bill Courtenay was there to greet her as she stepped down from the cockpit, smiling. She was wearing a leather helmet and a flying coat to match, which she quickly slipped out of to show herself dressed in a simple, sleeveless summer frock. It suited the occasion, for it was a perfect English summer's day with the thermometer in the mid-seventies.

One local newspaper reporter described the scene just before the 'intrepid girl flyer' arrived as reminiscent of a large race meeting. There were pedlars and hawkers selling souvenirs of the occasion with cries of, 'Amy Johnson brooches, sixpence each', and 'The only working model of Amy Johnson's aeroplane, just tuppence each'. Their cries competed with the roar of traffic and the amplified sound of gramophone music that was booming out from loudspeakers. Buses, cars, cyclists and pedestrians streamed towards the flying field, bringing young and old alike, all determined to see the girl who had captured the heart of a nation. It all began to look very much like an outbreak of Amy Johnson fever.

It says much for Amy's innate modesty that she apologised for her poor performance when she went up to the microphone and spoke to the crowd. 'Please do not judge by my rotten exhibition of landing. I am awfully sorry. I have had rather a strenuous time since I came back and I have had little chance to get back into flying trim.' Within minutes she was whisked off for another welcome at the Saffrons ground, where Sussex were playing cricket against Worcestershire, and then to the Grand Hotel for yet another civic reception. It was a slow, relentless grind of exposure for the 'Queen of the Air', one that was eventually to take its toll.

Bournemouth must have conjured up many poignant memories for Amy when she put *Jason* down at Purchase and Vine's farm at Wallisdown on the following afternoon. This time her landing appeared to pass without comment and she was escorted to Meyrick Park, where an estimated crowd of 25,000 were waiting to catch a glimpse of her as she opened Bournemouth's Hospital Fête. As she stepped up onto the verandah of the cricket pavilion to receive an official welcome, she was surprised to be greeted by her three Eddison cousins, Nancy, Jill and the irrepressible Bunty. She had not seen them since she was last in Bournemouth during her convalescence after the breakdown she had suffered in 1925. Bunty, her favourite, was now ten years old and no doubt full of admiration for her older cousin. There were affectionate hugs and she was obviously thrilled when Amy gave her some of the flowers from her bouquet.

Another nostalgic moment came soon after, when Amy realised that she had been booked in at the Royal Bath Hotel for the next two nights. It was here that she had scraped up enough money to spend a romantic evening dancing with Hans Arregger. At that time she was madly in love and imagined that she might soon be married. Then she had had a close confidant, someone with whom she could share her deepest thoughts. Now, although she was surrounded by ardent admirers, she was a woman totally alone in a crowd. She once told Jack Humphreys that she could never marry a 'Mr Amy Johnson', her ideal would have to be a man in his own right, one who was her equal in courage and daring. The difficulty was that she had set a high standard for any man to follow.

Bill Courtenay had arranged for Amy to make a mid-morning tour of the town by car on the next day, but he was forced at the last minute to announce that she was not well enough, since she was feeling the strain of her provincial tour. Instead she would be resting. The truth was that Amy had come to the conclusion that she was being pushed too far and she simply refused to go on. Although he was well intentioned, Courtenay was expecting too much from her. Instead, she left by the back entrance of the hotel and went for an early-morning swim in the sea. It meant disappointment for those lining the streets early to see her, but she now began to set her own agenda. She did, however, pull herself together sufficiently to fulfil the rest of the day's itinerary, which meant attending a concert at the Pavilion and a civic banquet given in her honour in the evening.

Amy's poor state of health had come to the notice of several people, even before she flew on to her next planned appearance, which was to be at Brighton on that final weekend in August. Lord Rothermere wired her whilst she was in Bournemouth to say that she was to take things easier and that she had done enough for the *Daily Mail* already in any case. It was almost a let-out from the contract for her. Obviously he did not want to be the one responsible if she had an accident. An added caution about continuing with the tour came from her father, who wrote pleading with her to be more careful, for he had read in the newspapers of the poor landing that she had made at Eastbourne.

Someone must also have alerted Brancker, since he now felt that he should be at Brighton for the Saturday evening to attend a dinner given in Amy's honour. No doubt his real intention was to advise her to give up the tour and take a complete rest, for on the Sunday evening she told Courtenay that this is what she wanted to do. Her decision caused an immediate rift between them. Courtenay told her that according to her contract *Jason* belonged to the newspaper, and that the *Daily Mail* was perfectly entitled to use their own pilot to fly the aircraft instead of her and to continue with the tour. This infuriated Amy, who told him that she would sooner burn the aircraft than allow anyone else to fly it.

Fearing that the newspaper could actually do what they had threatened, she took off on the Monday morning and flew her plane to Stag Lane, where she pleaded with Jack Humphreys to hide it away in one of the hangars. He agreed to do so, and she travelled into London to discuss with Associated Newspapers exactly where she now stood in the light of Lord Rothermere's willingness to allow her to back out of the contract. It appears that there was an amicable settlement whereby Amy retained the £10,000 as an outright gift, on the proviso that she fulfilled her obligations on the next two planned events. These would entail a personal appearance at Portsmouth and another at Ratcliffe, but in neither case would she be required to fly.

That weekend in Brighton was most probably the last time that Amy was ever to see Sir Sefton Brancker alive. Sadly, he perished, along with Lord Thomson and forty-six other members of the crew and passengers of the airship R101, when it crashed in flames at Beauvais on 5 October 1930 whilst on its maiden voyage to India. Amy might well have been one of the victims. Ciss Johnson, in a letter written to the mother of Amy's friends, Sheila and Mabel Glass, recorded:

> Amy has lost a great friend in Sir Sefton Brancker. She is very heart-broken for she knew so many of the officers. Amy begged Sir Sefton to get permission for her to go, but for some reason not known, no ladies were allowed . . . so she was terribly disappointed! Then again, the doctor took her in hand and compelled her to rest, no letters, no visitors, etc. If she had been a boy, she might have gone on the R101 and might have shared the same fate. 'IF'. So you can imagine my deep thankfulness.

After fulfilling the last of her contractual obligations by attending a rally of young people in the Portsmouth football stadium, and appearing at the opening of Leicestershire's new aerodrome at Ratcliffe during the first weekend of September, Amy took Brancker's advice to accept a holiday he had arranged for her with some wealthy admirers at Norton Priory, near Chichester. It was from here that she admitted to her parents that she was at the end of her tether and bordering on a nervous breakdown. We gain some further idea of her innermost thoughts at that time from a letter she wrote to Geoffrey de Havilland on 19 September:

> I'm not at all surprised you have refused invitations to dinners, etc., and I'm sure I'd do the same in your place. I loathe them myself and am also refusing them all. I'm afraid even those which have been accepted will be disappointed, because as soon as I have rested I hope to get abroad for several months.
>
> Publicity would in time drive me insane and I am therefore taking the cowardly action of running away from it. I'm sure you will understand how I feel . . . I'm looking forward to having my new Puss Moth and then I hope to be off on my travels again.
>
> There are not many people in the aviation world who understand my position and I feel that the unwanted and overdone publicity I've received has just about ruined any aviation career I might have had in this country. My one desire is is to be left in peace to fly.

Not everyone was sympathetic to Amy's reaction to the publicity she was receiving. Some at the London Aeroplane Club took the view that 'she sought it, and now she's got it she's not happy. It's her own fault.' They saw her as an accomplice in her own predicament. No doubt Amy was naive in many ways, for although her father managed her business affairs and Vernon Wood became her legal adviser, she still lacked a protective publicity agent when dealing with the news-hungry wolves in Fleet Street.

When one compares the way in which the flying career of her American counterpart, Amelia Earhart, was carefuly scripted and stage-managed by her astute and publicity-conscious husband, George Putnam, one can see quite clearly how much Amy lacked this kind of shield. Putnam would advise his wife down to the tiniest detail, even insisting that when photographed in a group that she should always stand at the end of a line on the right-hand side. This ensured that when her picture appeared in the press, the caption always began with his wife's name. And whenever Amelia was involved in a mishap which might have been due to a lapse in her own flying skills, her husband would be on hand with the press to put it down to a mechanical failure or something beyond her control. When she became the first woman to make a solo crossing of the North Atlantic, she landed in Northern Ireland. When she telephoned her husband to tell him of her safe arrival and also told him that she had received offers from British aviators, among them Amy Johnson and Lady Bailey, to fly her to the reception that was awaiting her in London, he advised her not to accept in case they should share in her limelight.

Amy's stay at Norton Priory lasted a mere ten days, although she could probably have stayed longer – and should have done so. But she was anxious to try out her new cabin-monoplane, the brand new Puss Moth given to her by de Havilland, which she christened *Jason II*. It was a big improvement on her old, open-cockpit Gipsy Moth biplane, having the advantage of increased speed and comfort, and the capacity to carry two passengers. It also had a much reduced noise level, an important factor on long-distance flights. Amy was anxious to show it off to her family and was able to fly up to Hedon for the weekend, where she gave pleasure flights to her parents and two sisters.

When she returned to Stag Lane the following week, Jack Humphreys must have been concerned for her mental health, since he persuaded her to visit a Harley Street physician, who warned her that she needed complete rest. Within days she was in a London nursing home, with strict instructions to make a break with the outside world. As we have seen from her mother's letter, there were to be no visitors and no letters to or from her.

Some have said that Amy was bordering on a complete mental breakdown at this time, but this would seem to be something of an exaggeration. She was, however, highly strung, which meant that she was given to uncontrollable outbursts of temper in certain situations. The pressure of her sudden fame was bound to exacerbate any natural tendency that she had towards instability. And ever since leaving the Boulevard she had been under considerable stress. First she had to cope with university whilst at the same time conducting an obsessive love affair. Then came the split with her family over Hans, with her leaving home to grapple with the loneliness of being a stranger in London. This was followed by rejection and humiliation at the hands of the man she loved, and finally having to come to terms with her sister's suicide. All of this, even before the flight to Australia, must have taken its toll. Small wonder she buckled. When today's celebrities come under similar pressures, they often find an escape in drugs – Amy just had her nervous breakdowns.

Never one to give in to a situation, she managed to persuade her doctor to allow her to attend what she considered to be a most important function, one she did not want to miss. The Society of Engineers had arranged a dinner in her honour for 21 October at the Holborn Restaurant in London, at which she was to give a paper on her flight to Australia, with particular emphasis on the way she had maintained her engine throughout the journey. The date had been carefully chosen because it was Trafalgar Day, a day commemorating Nelson's victory at sea, whilst Amy's had been in the air. By accepting their invitation and attending, she hoped that she would send out a signal to the industry that she was not just a stunt flyer, but a woman who was worthy of a career in aviation. She was desperate to be taken seriously.

The speech held her knowledgeable audience in rapt attention for almost an hour, as she spoke with clarity and without hesitation. If she was suffering from nerves then she made a very good job of concealing them. The large number present were impressed with her technical knowledge and the skill with which she put her points across, so much so that she was made an Honorary Fellow of the Society of Engineers, and later awarded the President's Gold Medal. No doubt there was none more full of admiration that evening than her mentor Jack Humphreys who, along with her proud parents, was present for the occasion.

Amy's audience appreciated that she had made two great strides for the emancipation of women. She had shown that her engineering knowledge had played a large part in her successful flight to Australia, and that both in the air and on the ground, women were capable of filling roles that had once been thought to be the sole province of men.

Chapter Eleven

CALL OF THE ORIENT

Many people have said that Amy had nine lives, and that one of them was used up in the late afternoon of 9 November 1930, when she set out to fly to Hull. It was only the fourth outing that Amy had made in her new Puss Moth, G-AAZV, *Jason* II, and she set off from Stag Lane in conditions that were far from ideal. Just north of the capital she ran into fairly thick fog which reduced her visibility to such an extent that there were only two courses of action open to her: she could either turn around and go back, or she could climb above the murk and hope that it was clear over her destination. She chose the latter course and climbed to 6000 feet, where she broke through the cloud layer and emerged into brilliant sunshine.

After flying on a compass course for the estimated time required to be over Hedon, she descended through the cloud, only to find that she was over the sea. She quickly realised that there had been a drift in her course for which she had not allowed. It was almost pitch dark and as she descended further to 500 feet, she was horrified to see nothing but the white caps of the waves of the North Sea. She flew around in ever-widening circles for some time in the hope that she would eventually see land, and was relieved when she finally saw what she imagined to be the twinkling lights of the shoreline. However, as she approached them she discovered that she was circling over an ocean liner.

She decided that her only course now was to ditch in the sea alongside the liner in the hope that she would be rescued. Just as she was contemplating slipping off her shoes and unlatching the door in preparation for a speedy exit, the fog lifted slightly, but enough for her to see another light further away in the distance. As she turned towards it she realised that it was the beacon of the lighthouse at Spurn Head. Within a few minutes she recognised the shoreline and was able to find the only piece of sandy beach available. After making one or two low passes to make sure that it was a suitable surface, she made a successful landing between two of the groynes. It was a very near thing and she breathed a sigh of relief as she got down from the cabin and made her way along the shore to seek help.

As she answered the phone at 85 Park Avenue, Mollie was surprised to hear Amy's voice, telling her what had happened and requesting that her mother come out and meet her with the car at Easington station. Ciss was away at the time and it meant that Mollie, who had only just learned to drive, was prompted to be the chauffeur. Meanwhile, the coastguards nearby needed little persuading to help once they realised that it was Amy who was in trouble, and the aircraft was pushed to a safe position away from the incoming tide, where a watch was kept on it overnight. The following morning Amy flew the machine off the beach to the aerodrome at Hedon and was able to spend a few days with her family before attending the Scarborough Gliding Fête.

The whole incident of the Spurn Head forced landing and its implications have been largely overlooked by those aviation historians who have, over the past sixty years, held entrenched views on the circumstances surrounding Amy's death in 1941. One needs to bear in mind the action that Amy was contemplating just before she saw the flashing beacon of the lighthouse for, as we shall see later, it has a special significance.

Ever since Amy had left home to seek employment in London three years earlier, she had lived a gipsy-like existence, moving from one lodging place or shared flat to another, never quite able to settle. Now, in December 1930, when she was able to purchase a flat at Vernon Court in Finchley, she found that it did not cure her restlessness. The chore of answering fan mail, together with the press coverage and the constant vigil of the paparazzi, was more than she could stomach. She longed to get away from living a fish-bowl existence and it was no doubt her eagerness to do so that warped her judgment on the timing of her next major solo flight which was to Peking.

Maybe she was impatient to prove to those in the aviation world who mattered, that her flight to Australia was not just a one-off publicity stunt by 'a petticoat pilot', a term once used by C. W. A. Scott to disparage her, but that she was someone who was to be taken seriously. However, right from the start she was advised against her choice of aircraft and the timing of the flight. For some unknown reason she decided to travel, not in the comparative comfort of her new Puss Moth, but in the open-cockpit Gipsy Moth which had been presented to her by the readers of the *Daily Sketch*.* Considering that she intended to make the journey early in the New Year, on a route that would take her across Europe and Siberia in the middle of winter in sub-zero temperatures, it was, to say the least, an ill-thought-out scheme. But, as Jack Humphreys discovered when he worked with her at Stag Lane, she could be headstrong, and then there was no way of talking her out of an idea.

Amy's plan was to fly via Berlin to Moscow, where she would exchange the wheeled undercarriage of her biplane for skis, and then follow the Trans-Siberian Railway via Kazan, Sverdlovsk, Omsk and Novosibirsk to Irkutsk, and then to Peking via Chita, Harbin and Mukden. The Soviet Union's civil aviation authority had already planned a series of primitive airports along the railway route linking Moscow with Vladivostock, and it was these that Amy hoped to use.

She took off from Stag Lane on New Year's Day 1931 and refuelled at Lympne with the intention of making Berlin that day in one hop. Instead, she was forced to spend the night at Liège and then the next night at Cologne, when it was reported that a band of fog stretched right across Germany. If she had been sensible she would have returned to the UK and delayed the flight until conditions improved, but presumably her pride was at stake and she pressed on. The weather was no better on the the third day when she left Berlin bound for Warsaw, and after five and a half hours in the air in freezing fog and temperatures below zero, she was forced to make a landing near the village of Amelin, some 55 miles north of the Polish capital. She picked what she thought to be the smoothest spot available on which to make a landing, only to discover that beneath its snow-covered surface lay a frozen potato field. She struggled to keep control of the machine as it lurched and bounced across the rutted field until it finally nosed over. It came to an ignominious halt as the propeller splintered into fragments and the undercarriage buckled.

By 7 January Amy had decided that there was now no possibility of her continuing the flight across Siberia under such conditions, and with the co-

---

* Amy's original aeroplane, *Jason*, G-AAAH, was bought for £300 by the *Daily Mail* in settlement of a breach of contract action brought against her. Much like Lindbergh's aeroplane, *Spirit of St Louis*, it was then donated to the nation. Within a short while it was placed in the Science Museum in London, where it remains to this day.

operation of the British Embassy in Warsaw she left her aircraft with the Polish Air Force for repair whilst she took a train to Moscow. She knew she would receive a warm welcome in the Soviet capital, for not only was the country air-minded, but women were also more readily accepted in aviation than in her own country. Even the head of the Russian equivalent of the Air League of the British Empire was a woman, and she not only befriended Amy but also arranged for her to be given a flight in one of their training aircraft. They obviously recognised in an international figure such as Amy, a propaganda opportunity to promote themselves as the equal of the Western European countries in aviation development. To this end they were only too willing to assist her in any future plans she might have which entailed flying across their territory. As a result, even before she left to collect her repaired aircraft, Amy was poring over maps supplied by her hosts and planning her next attempt on a flight to the Orient. She was determined not to be beaten – she would be back!

Within a few days of returning to her London flat Amy was pleasantly surprised to receive an invitation to travel to Switzerland and join the Glass family, who were spending a skiing holiday near Montreux. They had read the reports of Amy's mishap in Warsaw and had immediately urged her to join them. The two young daughters, Sheila and Mabel looked upon Amy very much as their big sister and through some secret family joke always referred to her affectionately as 'Mickey'.* Photographs taken at the time show a relaxed and smiling Amy dressed in tartan skirt, long-sleeved shirt and tie skating with Sheila; and another showing her sprawled on her back in the snow laughing, with her skis awry. The caption to the latter reads, ' "Sea Lion" Mickey after fall, Chosieres'.

It says much for Amy's integrity that throughout her life she was loyal to her friends. Always a prolific letter writer, she never failed to acknowledge even the most trivial of favours or kindnesses she received, and although she mixed easily with the rich and famous, she preferred to be with the flying fraternity and the lifelong friends she made at Stag Lane. Two of those friends were Pauline Gower and Dorothy Spicer, who were now running an air charter business. When Amy returned from Poland they were in need of a

---

* As a result of Amy's influence, the two sisters subsequently became accomplished flyers in their own right. Mabel Glass later served during the Second World War as a First Officer in the Air Transport Auxiliary.

further aeroplane and she came to their rescue by loaning them her Gipsy Moth.

Pauline Gower was the daughter of Sir Robert Gower, MP, and a natural pilot who had gained her 'A' and 'B' Licences at the London Aeroplane Club before taking her Second-Class Navigator's and GPO Wireless Licences. According to Eva Fitzpatrick she was also a skilled horsewoman who kept an elegant thoroughbred bay mare named Vogue, 'thus giving support to Travers's pet theory that good pilots often come from the ranks of good horsemen and women'. Dorothy Spicer, described by Amy as 'a very pretty girl with long, fair hair, a delicate white skin and rosy cheeks', qualified not only as a pilot, but also as a trained engine fitter and aircraft rigger. The two women subsequently teamed up and pooled their resources to buy a second-hand Spartan three-seater biplane in which they toured the country giving pleasure flights to the public.

There can be little doubt that Amy's epic flight to Australia inspired not only Pauline Gower and Dorothy Spicer to further their careers in aviation, but also another woman who went on to make a name for herself in long-distance record-breaking flights, namely the glamorous New Zealander, Jean Gardner Batten. She made her first solo at Stag Lane under the watchful eye of Major Travers in September 1930, and much like Amy before her, had problems mastering the technique of landing. Her biographer, Ian Mackersey, in his book *The Garbo of the Skies*, recorded a fellow pupil saying she had 'such terrific trouble with her landings'. On one occasion, he recalled, 'She crashed into a fence; fortunately she wasn't hurt, just a little bruised, but the pluck of the girl is astounding'.

It was not only the women who aspired to outshine Amy Johnson. On 31 March 1931 C. W. A. Scott set out from Lympne in an attempt to break Kingsford Smith's solo record* from England to Australia. Scott landed at Darwin after nine days and four hours knocking almost seventeen hours off the Australian's record. He was challenged to make the return flight straight away and in a moment of bravado agreed. However, he then ducked out.

> My nervous and physical qualities were impaired, there was no doubt about it, and I contemplated this return flight with great misgivings, for anyone who has not had the experience of a long fast flight will hardly realise the strain and the natural reactions. Had I been allowed to rest after my arrival in Australia I should have been in a far better condition, but all these parties and evenings had given me no rest at all, so I insisted

---

* Kingsford Smith set a new England–Australia record of nine and three quarter days when he flew his Avro Avian, IVA, *Southern Cross Junior* to land at Darwin on 19 October 1930.

that I be left alone to prepare myself physically for this homeward flight.[12]

Scott's account must have brought a wry smile to Amy's face when she read it in 1934. The biter was bitten. So now he knew!

At the time of Scott's flight, Australian National Airways was going through a bad patch and the future looked uncertain for its pilots. One of their airliners, *Southern Cloud*, had disappeared with eight passengers and crew on board whilst on a scheduled flight from Sydney to Melbourne. Extensive searches had been made for the aircraft but without success, and after ten days the search was called off. The mystery of the airliner's fate continued for many years and was not solved until twenty-seven years later, when a trekker on a climbing expedition stumbled upon the wreckage and the remains of its victims in wild and mountainous country.

A court of enquiry was set up within a few weeks of the plane's disappearance and although ANA were cleared of any blame, the cost of the search and the ruling that the airline could no longer operate until effective air-to-ground communications systems were installed in their aircraft brought it to the point of bankruptcy. By the end of May 1931, Jim Mollison knew that his days with ANA were numbered.

He reasoned that if Amy Johnson, Kingsford Smith and Scott could pick up prize money by making record-breaking long-distance solo flights, then so could he. With almost 4000 hours to his credit in military and civil flying, he knew that he had all the qualifications to succeed. Moreover, if he could master the atrocious weather conditions on the Sydney–Melbourne run, which had killed Travers Shortridge and his passengers, he reasoned that he was more than capable of dealing with the worst that was likely to be encountered on a flight to England. There was just one snag. Where would the finance come from?

The answer came when he happened to meet Wakefield's manager, Cyril Westcott, and told him that he intended having a crack at the Australia–England solo record. Westcott was quietly noncommittal at the time, but at their next meeting a few days later, he told him that he had asked Lord Wakefield if he would help. No one was more surprised than Mollison when, some weeks later, Westcott approached him with a cable from Wakefield . On it were the three brief words, 'WILL BACK MOLLISON'.

Amy returned from her holiday in Switzerland refreshed and determined to restore the damage she had done to her public image by her ill-advised attempt to fly solo to Peking. Fortunately, her mishap at Warsaw seemed to attract little attention from the media, and she was soon making plans for a further flight to the Far East later in the year. An added incentive to do so came when Amy and Winifred Spooner were invited to act as welcoming air-escorts to another woman pilot, the Hon. Mrs Victor Bruce, when she arrived back in the UK in February after making a 20,000-mile flight around the world in her small blue and silver biplane. Her open-cockpit Blackburn machine was registered as G-ABDS, which prompted the more cynical to dub it 'a bloody daft stunt'.

It had taken the plucky Mildred Bruce five months to accomplish her circumnavigation of the globe, and in spite of the fact that both the Tokyo to Vancouver leg and the Atlantic crossing had been made by ocean liner, it was still credited with being the longest flight ever made in a light aeroplane. The final stage of her journey, however, had been from Le Havre to Lympne, where she was joined by her two escorts to accompany her to Croydon. Much to the amazement of the crowd and waiting reporters, she stepped from the cockpit smiling, dressed in high heels, blue jumper and grey skirt, as if to go shopping down the High Street. Her small son was lifted up to kiss her and was heard to whisper, 'Mummy, I've won a prize in Geography.' To which his mother replied, 'Well done, my flight round the world has done some good anyway.'

The reports of the extraordinary welcome that Mrs Bruce received in Tokyo might well have influenced Amy in her decision to make the Japanese capital her target for her record-making Far East flight, and not Peking. A further factor in her choice was that the Japanese were in dispute with China over an alleged infringement of their interests in the South Manchurian Railway and, in spite of efforts by the League of Nations to intervene, had invaded Manchuria and parts of China, even as Amy was planning the flight.

This time Amy decided it would be more prudent to use the Puss Moth in its standard trim and to take Jack Humphreys along with her to act as co-pilot and mechanic. This meant she was unable to carry a long-range fuel tank in the cabin, and as a consequence she would have to put down more frequently for refuelling. However, she reasoned that this disadvantage would be more than offset by being able to demonstrate just what a standard aeroplane could achieve on a long-distance flight.

Jack Humphreys jumped at Amy's offer; since breaking his right arm when one of the club's machines had overrun its chocks, he had only been on light duties with the club. Whilst many people only ever saw him as a ground engineer, he had in fact served as a pilot during the First World War, and had seen action in France with the Royal Flying Corps. One colleague who worked with him during the post-war years recalled that, although he rarely spoke of his wartime experiences, he could, during a slack spell in the hangar, be coaxed into saying something about of his remarkable RFC exploits.

On one occasion he told how he was on a routine reconnaissance flight when an enemy fighter made a brief passing attack from above and behind, and then immediately broke away. Before he had time to take evasive action, a hail of bullets struck the cockpit and he felt a numbing pain in his feet, both of which were on the rudder bar. He managed to get back to his base, where he discovered, as he was helped down from his machine, that both his flying boots were filled with blood – he had lost the second toe on each foot. At this point in the story Jack would respond to the look of incredulity on the faces of his listeners by saying, 'You don't believe me now, do you?' Whereupon he would kick off his shoes and invite them to feel the two gaps where his toes should have been.[13]

Shortly after midnight on 28 July 1931 Amy Johnson and Jack Humphreys took off from Lympne in *Jason* II on their way to Tokyo. Their departure attracted negligible press coverage, which is exactly how Amy preferred it to be, since she was still smarting from the embarrassment of her abortive attempt to reach Peking earlier in the year. The Puss Moth's two upper wing tanks carried a mere 40 gallons, which meant that they were obliged to put down at Berlin, Konigsberg and Velikije Luki on the Russo-Latvian border before they made Moscow. However, the fuel stops allowed them to share the flying, one at the controls whilst the other attempted to take a nap, thus enabling them to cover a creditable 1760 miles in a single day. It was the first time that Moscow had been reached in a day from England, and as William Courtenay observed, 'The long stage gave her invaluable experience for her future record flights.'

Over the next six days they followed the Trans-Siberian Railway, often losing sight of the line as it passed through dense pine forests. Jack Humphreys recalled that on at least one occasion he was surprised by Amy's acute sense of direction when they were uncertain of their exact position. Of her landing approaches, he was much less complimentary. However, as a team they worked well together. One of the main benefits was that whereas on her Australia flight she was required to spend precious time at the end of

Amy and Jack Humphreys alongside the DH Puss Moth, Jason II *(G-AAZV) just prior to their departure for Tokyo. A nice touch is added by the High Street carrier bag – as if to show that a 14,000 mile flight was nothing unusual in 1931.* (Hull Daily Mail)

each day carrying out her servicing routine, she was now able to catch up on lost sleep while he did the work for her. Their overnight stops along the way were made at Kazan, Kurgan, Tiajin, Irkutsk, Chita, and Hailar as they passed into Manchuria.

Amy was anxious not to become involved with the authorities in Manchuria, since the Japanese were now in control of the area and highly sensitive to the possibility of any aerial photography which might reveal military movements and installations. Many years later an admirer and collector of Amy Johnson memorabilia managed to purchase a photograph album, allegedly rescued from the loft of her sister Mollie's former home in Blackpool, which show this particular period in the life of the aviatrix. Not only are there photographs of Amy and Jack alongside welcoming groups of British expatriates at Harbin, but also of Japanese troops marching triumphantly through one of the neighbouring villages, proudly carrying the unfurled flag of the rising sun.

The couple reached Tokyo via Mukden, Seoul and a 350-mile crossing of the Sea of Japan just ten days after leaving England. It was the first direct flight from the UK, and improved on an existing record by a Japanese pilot from Berlin to Tokyo by more than a day. They were given a rapturous welcome in the capital and spent a week there whilst Jack overhauled the engine, ready for the flight home. There were no great rewards for their achievement, other than the satisfaction of knowing that the trip had restored Amy's confidence and given her the opportunity to sharpen up her navigational skills.

Although Amy and Jack had made a record-breaking trip of 14,000 miles from start to finish, it received scant attention from the media back home. There were two reasons for her failure to capture the headlines. The first was that the timing of her flight was unfortunate inasmuch as just as few weeks earlier, the one-eyed American pilot, Wiley Post, and his Tasmanian navigator, Harold Gatty, had amazed the world by circumnavigating the globe in their Lockheed Vega, *Winnie Mae*, in just over eight days. Admittedly their flight fell short of an equatorial distance by some 10,000 miles, having been flown at a northerly latitude of between 45 and 60 degrees, but it still continued to outshine Amy's achievement. The second reason was that Jim Mollison, with the backing of Lord Wakefield, had made an outstanding solo flight in a Gipsy Moth from Australia in just under nine days and had arrived in England on the very day that Amy arrived in Tokyo. He had set a new Australia–UK record by knocking more than two days off C. W. A. Scott's time.

Upon her return Amy set off with Jack Humphreys on a lecture tour of the country, but not before instructing him to search out a suitable aeroplane for a round-the-world flight. It seems that Post and Gatty's success had inspired her to emulate them, and approaches were made to the Vickers Aircraft Company at Weybridge for the use of one of their Viastra twin-engined monoplanes. Amy let it be known that she proposed to make the flight during the following summer; however, nothing ever really came of her plans and the idea was dropped.

One question that must have exercised the imagination of many of Amy's admirers concerned whom she might marry. After her affair with Hans, her career had taken off, and her fame had now isolated her from the very men to whom she might have been drawn. She was attractive and desirable, as her fan-mail shows. However, one can well imagine that men would find it difficult to take the romantic lead in approaching a woman with such iconic status.

In spite of her close relationship with Jack Humphreys, there does not appear to have been any sexual attraction between them. He was after all a married man who was merely content to share in her successes and to bask in the reflected glory of someone he deeply admired. This did not mean that he was blind to her faults, since he was all too aware that she could be a difficult person to cope with emotionally, especially when she was in what he termed 'one of her black moods'.

Amy's name has been romantically linked with one of her more ardent admirers, Peter Q. Reiss, a young aviation underwriter working for Lloyd's. She had met PQ, as he was known to all his friends, through her contact with the Leicestershire Flying Club at Ratcliffe, where he acted as private pilot to Sir Lindsay Everard. Whilst he had acquired a reputation as a skilful pilot during his wartime service with the RFC, and was also a champion squash player, as far as Amy was concerned there were no reciprocal feelings towards his advances. Not only was he already married, and in those days young women in the public eye just did not play around with married men, but as later correspondence between Amy and her father reveals, whilst she accepted the help and advice that he gave her on her later flights, as far as romance was concerned he could forget it. The Queen of the Air had yet to find her consort.

Chapter Twelve

CONSORT FOR A QUEEN

im Mollison's reception at Croydon on 6 August 1931 at the end of his record-breaking solo flight from Australia was unusual to say the least. Amongst the welcoming crowd that afternoon was an unknown Australian who, for some obscure reason, believed that Mollison was also native to his own country, and had thought it appropriate to bring along with him a real live boxing kangaroo. The fun first began when the animal broke loose from its owner and came perilously close to becoming minced in the whirring propeller of the Gipsy Moth as it landed. And then a few minutes later, as the bleary-eyed Scot walked towards the official welcoming party, the animal kicked him smartly in the stomach. Such bizarre happenings almost caused the whole event to degenerate into pure theatrical farce, although from the photographs taken at the time, Mollison appears quite unfazed by the whole affair. No doubt with his pawky sense of humour, he would have had a few choice words to say to this Australian admirer.

After making himself presentable, Mollison was escorted to the nearby Aerodrome Hotel where a luncheon was given in his honour. He remembered very little of its proceedings however, for after consuming the better part of a bottle of champagne, he literally fell asleep whilst on his feet and in mid-sentence during his speech of reply. The next thing he recalls was waking up in one of the hotel's bedrooms, where he had been carried to sleep off some of his tiredness.

Telegrams of congratulation began to arrive, including one from King George V, as well as invitations to attend and speak at various functions. One phone call was from the Chairman of the Grosvenor House Hotel in Park Lane, offering him the complimentary use of a suite of rooms, much as had been done for Amy a year earlier. Another call came from a tailoring establishment in London, insisting on supplying him with any clothing he might need. Slowly, it dawned on him that he had become a celebrity overnight. He had arrived back in the UK virtually penniless, possessing nothing more than four £1 notes in his pocket and the Gipsy Moth that now stood in the hangar. He had stepped across an invisible line. On one side obscurity; on the other side, instant fame.

There can be no doubt that Mollison was flattered when he received Amy's cable of congratulation from Tokyo. If nothing else it meant that she had not forgotten the dashing young airline pilot who had chatted her up in the cockpit of the *Southern Sun* only the previous year. It was also obvious that she recognised him as someone like herself, one who was prepared to risk his neck in order to achieve a goal in life.

It has been estimated that Jim Mollison collected approximately £7000 in prize money for his Australia–UK flight. This sum included a hefty cheque from Lord Wakefield, together with a £1000 gift from the Australian philanthropist, A. E. Whitelaw; plus money from the *Daily Mail* in exchange for a contract to make an instructional flying tour of Scotland. In addition, there were the perks that he would pick up for broadcasts, public speaking engagements, the use of his name in advertising and the gifts which were showered upon him.

In an age before television, these early air pioneers and long-distance solo aviators, men and women such as Charles Lindbergh, Wiley Post, Jim Mollison, Amelia Earhart and Amy Johnson, were more than the equivalent of today's superstars. They vied in popularity with the idols of the silver screen by virtue of the fact that the golden ages of aviation and Hollywood coincided. It was not only Rudolf Valentino whom girls on both sides of the Atlantic swooned over; or only Clara Bow who caused men's hearts to beat that little bit faster. As air historian Richard K. Smith pointed out, this was very much the case when Charles Lindbergh was given a reception in New York soon after his North Atlantic crossing. He recorded:

> Women literally flung themselves at him, sometimes singly, sometimes *en masse*. He once accepted an invitation to a privileged club on Long Island, thinking he would be among well-mannered people, but he found the young matrons and unmarried women panting after him even harder than the girls in the street.

These new heroes and heroines of the twenties and thirties were treated as demigods. They were the last of the old frontiersmen and women. From now on outstanding achievements such as theirs would only be accomplished by teamwork and not by lone individuals.

In his memoirs, *Playboy of the Air*, Mollison told how he was introduced into a society which he describes as 'this world of film-stars, politicians, royalty and uncaught financiers'. West End society was always on the look-out for a new acquisition, a new playmate, and the young Scot who was now being lionised in the capital was only too willing to participate. Mayfair, that notorious hotbed of hedonism which falls roughly within an area bounded by Picadilly, Regent Street, Oxford Street and Park Lane, became his happy playground.

Tall, statuesque Kathleen, Countess of Drogheda, was mainly reponsible for introducing him to the rich and wealthy in London society, especially the pleasure-seeking side of it. She had a penchant for being seen in the company of men who were renowned for their bravery, particularly if it involved speed. One imagines that for her it was a kind of aphrodisiac – a turn-on. Her silver Rolls was frequently seen parked outside the nightclub of the moment, where she could, over a drink or two, accurately rattle off the names and family gossip of most of its denizens as they arrived, even as late as four in the morning. Drinking in properly licensed clubs could continue until 2 a.m., but these restrictions were usually circumvented by the means of 'bottle parties', where guests brought their own drink with them. Some of the more seedy nightclubs were run much on the lines of the American speakeasies during Prohibition, with liquor served in teacups, just in case the police made a raid.

The social circle in which Mollison's obliging chaperon moved fascinated and enthralled him, much like a boy with a new train set. He came to love the crowded, smoke-filled basements, where quick-fire chatter competed with the noise of the music and the cocktail shaker. These were clubs where he could gyrate to the syncopated sound of small jazz groups. He loved to be amongst Mayfair's 'bright young things' as they crowded onto postage-stamp-sized dance floors, shuffling to the mellow sounds of clarinet and saxophone. The pink champagne would flow until they were legless, and nobody cared.

One of the more respectable establishments was the Embassy Club in Old Bond Street, where one could dine and dance to top-class bands such as those led by Bert Ambrose and Jack Harris. It was here that the smart set could rub shoulders with royalty, for one of its regular patrons was the Prince of Wales, a man renowned for his love of the nightlife in London's West End. As one would expect, the golden-haired, 'little-boy' prince was never short of a dancing partner. According to one titled lady it was not only his rank and eligibility for marriage which impressed the ladies, but 'the intense blue of his eyes and the curiously wistful expression in them'.

Kathleen moved in the same social circle as the future monarch, and it was most probably through her influence that Jim Mollison received a personal invitation to meet the Prince of Wales for drinks at St James's Palace. The Prince took a great interest in aviation for he and his younger brother, the Duke of York, had both received flying instruction with the RAF at Croydon in 1919. Although the younger of the two princes qualified and received his wings, at the last moment the King intervened and withheld his permission for his eldest son to go solo. The decision was made on the grounds that, being direct in line for the throne, it was too risky for him to fly alone. As one destined to become the future king there was little he could do but accept his father's wishes. However, a meeting with Lindbergh soon after his Atlantic crossing inspired the prince to purchase his own de Havilland Moth aeroplane and to employ the RAF officer, Squadron Leader Don, to act clandestinely as his flying instructor. In spite of his father's protestations the determined prince went on with the flying lessons, and with the connivance of Don, even made an illicit solo flight long before he met Mollison.

Any nervousness that Mollison might have felt in his first meeting with royalty was soon dispelled as gin and tonics were poured and the two men began to discuss the finer points of the Australian flight. Although in his late thirties, the successor to the throne always seemed to retain a certain boyishness, which tended to reinforce his image as the world's Prince Charming. Only the premature pouches under his eyes betrayed his love of regular partying and all-night revelling. (The King was once quoted as saying, 'I don't know what will become of the boy when I'm gone.' The 'boy' was then almost forty years of age!)

Maybe the prince and Mollison recognised that they were in some way kindred spirits when it came to affairs of the heart. Both had a capacity for falling in and out of love very easily. At the time of their meeting, the prince was in the cooling-off stage of a long-standing liaison with Freda Dudley Ward, which also paralleled an affair with the desirable and titled Thelma Furness (Thelma's friend and fellow-American, Wallis Simpson, was yet to appear on the scene). All three women were already married, which is all the more remarkable when one remembers that the future king could have had the pick of any number of single and eligible women throughout Europe.

It was in the early part of 1932 that the countess almost certainly introduced her protégé to Lady Diana Wellesley, the great grand-daughter of the Duke of Wellington. Mollison describes meeting this 'charming and beautifully gowned eighteen-year-old', and how they quickly became secretly engaged. It was not long before the gossip columnists were reporting that they were being seen regularly dining and dancing together in the capital's night spots. With his good looks and reputation as a daredevil adventurer, Jim Mollison had little difficulty in attracting the right kind of women. His cut-glass accent and persuasive charm combined to cause many a young girl's heart to flutter. Although it swept Diana off her feet, it made little impression on her stepmother, Lady Clare of Cowley, who frowned upon what she perceived to be a *mésalliance* right from the start. Her opinion was strengthened when she read in the press that Mollison had been charged at the Marylebone Police Court with an assault on a member of the public.

Evidently, Mollison and two women had been standing outside a private residence in Bayswater waiting for a taxi, when the owner of the property came out of the house and ordered them to move away. According to Mollison the man verbally abused the two women, thinking that they were prostitutes and a fight had ensued. Immediately the police were called and the two women fled, obviously not wanting to be identified with the fracas. When asked later in court to name the two women, Mollison was only prepared to write their names down on a slip of paper, which was then passed to the magistrate. At least it showed him in a chivalrous light for it was reported in the press that one was a titled married lady (presumably the Countess of Drogheda) and the other 'a widow'.

Although the case against him was dismissed, the press account did little to enhance his image in the eyes of Lady Cowley, and only served to harden her attitude against him. Her solution was to whisk Diana and her younger sister Cecilia off to the south of France on a motoring holiday. Mollison saw this as a direct snub. It made him all the more determined to succeed in a flight he was already planning. He hoped to capture the UK–Cape record, which had been set a few months earlier by Gordon Store and co-pilot Peggy Salaman. He was prepared to show that it was not only the old Iron Duke who could put up a fight. The would-be suitor for the hand of Lady Cowley's stepdaughter intended to be back in the newspapers, but this time with a success story.

By a strange coincidence Mollison arrived at Lympne on 19 March ready to make his attempt on the Cape record, only to find his old arch-rival

Charles Scott also at the Kent aerodrome. Scott was attempting to regain the UK–Australia record, which had recently been taken from him by C.A. Butler flying a tiny Comper Swift monoplane. Scott and Mollison had both picked the 20th as their departure date, solely because it coincided with a full moon.

In his memoirs, *Scott's Book*, the author recalled that the Mollison and Lady Wellesley affair was very much in the news at that time. He wrote:

> The following morning the placards in the streets of Hythe bore the words, 'Famous Airman's Romance', and I bought a newspaper and showed it to Mollison at breakfast as he was the airman concerned.
> Mollison left Hythe and stayed mysteriously away all day, and in spite of bad weather conditions prevailing, left for Cape Town the following morning. Publicity of this romance of his was too much for him, and he sought the security of the air and escaped.[14]

After spending the Christmas of 1931 with her parents at their new home in Bridlington, Amy resumed her lecture tour of the north of England with Jack Humphreys. She was giving illustrated talks on her Australia flight which were advertised as 'How Jason and I Flew to the Land of the Golden Fleece'. It was quite a lucrative undertaking, since she was getting a fee of 30 guineas for each talk she gave. Even if she only did two in a week, it was enough to enable her to

An advertisement appearing in the Blackpool Evening Gazette on 26 September 1931 announcing Amy Johnson's lecture, 'How "Jason" and I flew to the Land of the Golden Fleece.'

live in style and still be able to pay Jack more than he had earned at Stag Lane.

However, the lecture tour did not preclude Amy from accepting other invitations, one such being to her old school at the Boulevard, early in the New Year. Margaret Holliday was an eleven-year-old schoolgirl at the time and she can remember quite vividly the excitement amongst both staff and pupils when their heroine arrived. The children lined the staircase as Amy entered, and Margaret recalled that it was the first time that she had ever seen a woman wearing lipstick. She can also remember nudging her school friend as Amy passed by and whispering, 'Whatever will Miss Sheppard say?'

The lecture tour came to an abrupt end early in February, when Amy was suddenly taken ill with severe stomach pains whilst driving back from Bolton with Jack Humphreys. It meant Jack taking over the wheel and driving down to London through thick fog and with Amy lying prostrate in the back of the car. She had suffered pain for many years with irregular periods and had already discussed the problem with her doctor, who now ordered her into the Duchess Nursing Home in Beaumont Street, W1., where she underwent an immediate hysterectomy. The press came out with the official and more discreet version of her trouble by reporting that she had been operated on for an appendicitis. It now meant that she would never be able to have children, something she had desperately wanted at one time. The only consolation she could draw on was that she would no longer suffer as she had done on her long-distance flights.

Ciss came down to London to be near her daughter during the three weeks that she was in the nursing home. Writing to her friend 'Mrs Glass' (how formal people were in those days!) Amy's mother told her:

> Dear Johnnie has been very ill indeed . . . and was operated on for an appendicitis. You would have seen it in the newspapers, I think? Am delighted to say she has made a marvellous recovery . . . and will go into Essex [a convalescent home] with a nurse for a fortnight, after which she is contemplating a trip on the Mediterranean to get some sunshine. All the lectures have had to be cancelled – she may give them in the autumn. I think she will be far better in health afterwards, for she always looked so thin and tired.

Amy, as impetuous as ever, stuck the convalescent home in Colchester for only a few days before she pestered her doctor to allow her to make a trip to Madeira, where she hoped to soak up some sunshine. He agreed and she boarded the Union-Castle liner *Winchester Castle* at Southampton, along with others rich enough to escape the worst of the winter climate.

The lavender-hulled liner nosed into the small harbour at Funchal under grey, threatening skies and Amy disembarked with her luggage to take a taxi up along the steep cobbled streets to the Reid's Palace Hotel. It did not take her long to decide to check out of the five-star hotel, once she discovered that it had been raining in Madeira for the past six weeks. The prospect of sitting around in the hotel lounge amongst a gossipy group of society folk and becoming the object of curiosity, was not exactly her idea of a holiday. She immediately re-embarked on the *Winchester Castle* and booked an onward passage to Cape Town.

A day out of Cape Town, Amy received the news over the ship's radio that Jim Mollison was shortly due to arrive in South Africa after what looked to be a record-breaking flight. It happened to be the same day that the ship docked. There is no way of knowing if Amy had any romantic feelings running through her mind at that time. She had only met the Scot briefly two years earlier and since then, apart from the telegram of congratulation she had sent him when he arrived at Croydon, there had been no communication between them. There is no doubt that he fitted one part of her image of the ideal man, inasmuch as he had proved himself to be as determined and courageous as she herself.

Mollison had not only flown a distance of 6255 miles in four days and seventeen hours, beating Store and Salaman's record by almost fifteen hours, but he had also become the first person to solo the Sahara on a flight to the Cape. Much like his flight from Australia, it was a great feat of human endurance, for he had only had a total of eight or nine hours' sleep during the flight. His achievement was to be widely acclaimed as a superb piece of navigation over the particularly hazardous western route. Moreover, it firmly established him in the front rank of the world's aviators.

Amy wanted to be one of the first to congratulate her fellow countryman when he landed, and travelled from her hotel to await his arrival at Wingfield's Cape Town aerodrome. His aeroplane was seen over the airport at 8 p.m. but then headed off across the city and out of sight. Mollison confesses that his eyes were now beginning to play tricks on him and he had duplicated vision. One can only guess at this point whether this was caused purely by fatigue or, as one suspects, by too much alcohol on an empty stomach during the last stages of his flight.

According to Mollison's own account, the lights of the airport only served to confuse him. He knew instinctively that he would never make a safe landing at Wingfield that evening, and as a consequence he turned to look for a suitable spot on the beach at Milnerton, approximately 5 miles away. Keeping to the seashore he finally managed to scrape down on the beach after narrowly missing the roof of a taxi-cab parked on the adjoining road. Evidently the beach had a far greater slope than he had bargained for and as he attempted to keep directional control the machine veered and ran into the sea to a depth of 4 or 5 feet before turning over. He immediately kicked out one of the cabin windows and fell into the sea, where he took in mouthfuls of sea water. Soaked to the skin he staggered up the beach to safety.

The astonished driver, who had been sitting in the parked taxi and had witnessed the whole scene, ran down onto the beach to assist the wet and bedraggled pilot. Within minutes Mollison was being whisked back to the airport, where Amy was amongst the large crowd awaiting his arrival. Just what she thought when she saw the exhausted and dishevelled figure that walked into the airport on that Monday evening is not known. He can hardly have been in a fit state to hold a coherent conversation with her, let alone convey any thoughts of a future romance.

One must not forget that if Amy had been reading any of the current English newspapers, then she would have known that they were reporting that Mollison was engaged to be married to Lady Diana Wellesley. Certainly the London *Evening News* had been printing articles by Bill Courtenay giving the Mollison 'romance' a high profile. However, when questioned on the affair by the Cape Town reporters the Scot was tight-lipped, and if anything seemed to be backing away from the possibility of a commitment.

We do know that on the morning after Jim Mollison's arrival in South Africa Amy spent time helping him to open the many telegrams of congratulation, and that during the afternoon they lunched together in one of the city's fashionable restaurants. Inevitably, the story quickly spread to the effect that they were planning a flight across the Atlantic together. Such rumours may have originated from Mollison telling the press of his future plans for an east-to-west solo crossing of the North Atlantic, or of Amy mentioning her intention to make a round the world flight with Jack Humphreys.

Amy sailed that same afternoon on board the *Winchester Castle* for Durban, having now dispensed with the services of her nurse. Obviously, she had decided that she was still in need of further recuperation from her recent operation. Some say that she suffered a relapse in health in Cape Town and was on the point of collapse when she reached Durban, but this is hardly borne out by her own account written a few years later.

> Once I passed through [South Africa] on a cruise I was taking for my
> health, visiting Port Elizabeth with its world famous Snake Park and on to

Durban, where I spent one of the happiest fortnights of my life at the home of a well-known big-game hunter, William A. Campbell, affectionately known as 'Wac' to all his friends. His lovely modern house stands on top of the cliffs,* sprayed by the breaking waves of the Indian Ocean which scatter their foam just the same on picturesque African kraals further along the cliff-side.[15]

Amy in Cape Town in March 1932 when she met Jim Mollison for the second time.
(Smith's Photo Agency)

Amy Johnson and Jim Mollison soon discovered that fame is a key which opens many doors. The mega-rich have always wanted to bask in the reflected glory of their not-so-rich but famous guests, and neither Amy nor Jim were able to resist their blandishments. The Scot was befriended in Cape Town by Sir Abe Bailey, one of South Africa's wealthiest philanthropists, and husband of Lady Mary, the well-known aviatrix. Another who smoothed

* Nganlana, Mount Edgecombe.

Mollison's stay in the Province was Sir Pierre van Ryneveld, himself a pioneer aviator and Director of Union Air Services. In his opinion Mollison's flight was 'as great a performance as one can imagine'. High acclaim indeed from one who knew the game from first-hand experience. (The two South Africans, van Ryneveld and Quintin Brand, had made the first flight from England to South Africa, when, in 1920, they covered the distance from Brooklands to Cape Town in forty-five days flying a First World War Vickers Vimy bomber.)

Mollison arrived back at Southampton on board the *Carnarvon Castle*, together with his battered Puss Moth, with two thoughts uppermost in his mind: to straighten out his future relationship with Diana and, equally as important in his eyes, to be the first person to solo the North Atlantic in the more dangerous east–west direction. Lindbergh had made it across solo with the assistance of the prevailing winds; no one had successfully made it alone the other way.

Jim Mollison's daring flight across western Africa to the Cape made him even more desirable in the eyes of the besotted Diana Wellesley. He was now well and truly the young girl's knight in shining armour, and her family's opposition to their engagement only seemed to have made her all the more determined to continue seeing him. However, her stepmother, Lady Cowley, insisted on meeting the man whom her stepdaughter intended to marry, and he was invited to visit the family home at Seagrey Manor, near Chippenham, forthwith. In the Scot's mind the prospect of encountering the haughty, titian-haired tigress probably required as much courage as confronting the uncharted sands of the Sahara desert.

JM flew down to the family residence six days after he arrived home and landed his newly restored aeroplane in a nearby field, where an anxious Diana was waiting to lead him to meet Lady Cowley. Diana waited discreetly outside in an adjoining anteroom, whilst her inamorato entered for an interview that turned out to be every bit as unnerving as any he had experienced.

Lady Clare was completely underwhelmed by the aviator's flying achievements and made it quite plain that she did not take kindly to his infatuation with her stepdaughter. Moreover, she intimated that his background and family pedigree did not measure up to the tradition and breeding that she would deem fitting for any proposed marriage. According to the Scot, he was given a third-degree grilling worthy of a New Jersey police chief suffering from a hangover. It was one which ranged from his present and prospective financial position to details of the most personal nature. The outcome was that he came away highly dispirited. It was reminiscent of how he had felt when rebuffed by Amy's chaperon in Sydney whilst attempting to further a relationship with her. The only concession that he could wring from his inquisitor was that marriage was not on the agenda until Diana was twenty. Diana, it appears, was almost prepared to take the plunge and elope, but not quite. The outcome was that they agreed to keep their engagement on trial for a year and wait to see if parental opposition softened.

Amy, looking tanned and fit after her sea cruise, arrived back at Southampton early in May, and was met on the quayside by her mother and Jack Humphreys. Her former flying partner said that he immediately sensed an undefinable change in her. The animated way in which she greeted them caused him to suspect that she was hiding some inner excitement which refused to be masked. He did not have to wait long before the mystery was solved and his suspicions were confirmed. The following weekend, when Amy and he were returning from an aeronautical trade display in Antwerp, she insisted on calling in at Heston before flying back to Stag Lane. He must have wondered why she broke the journey so near to home, but as soon as they landed he knew. Jim Mollison was waiting there to meet them.

Events moved swiftly after the Heston meeting, because Jim Mollison invited Amy to lunch the next day at Quaglino's, one of London's classier restaurants. He was surprised by the startling change in her appearance from the woman he met briefly in Australia and again in South Africa. She was no longer the tomboy who had sat beside him in the cockpit of the *Southern Sun* on the flight from Brisbane. Nor was she the wan, tired-looking girl he had met in Cape Town recuperating from her operation. Sitting opposite him now, with those seductive blue-grey eyes, was an elegant, self-assured woman. She was, he thought, the epitome of sophistication, and there seems little doubt that he was dazzled by her.

Their conversation during that meal must have included his affair with Diana Wellesley, for Amy would most certainly have known of their relationship, and now with his heightened interest in Amy, she would want to know the state of play. This is hinted at in his memoirs, where he said, 'We discussed women, aeroplanes and travel'. And so it was that over post-prandial liqueurs, he asked her to marry him.

One can only speculate on the accuracy of the 1942 Herbert Wilcox film, *They Flew Alone*, where it portrays the Quaglino's restaurant scene. The Yorkshire-born actress Anna Neagle, who knew Amy quite well before the Second World War, played the part alongside a rather over-camped Robert Newton, who portrayed a perpetually inebriated version of the man she was to marry. It seems that Newton's major qualification for the role was that he too was renowned for being a bit of a hell-raiser, as well as an alcoholic, in real life.

The table conversation in the film has the Scot asking Amy, 'Do you like me?', followed by a slight pause before he adds, 'Why don't you marry me and find out?' To which Amy replies with a smile, 'I'll take a chance.'

Chapter Thirteen

TEA FOR TWO

T here has long been a debate as to whether the marriage between Amy Johnson and Jim Mollison was based upon real love; there are those who accuse him of using her purely for his own self-promotion. Others believe that the marriage was suggested and conspired by Lord Wakefield as a business arrangement that would further their careers in aviation, and where by if they promoted his products he would supply their needs.

One must acknowledge that Amy had far more celebrity value throughout her life than her husband ever did. She became an icon, a symbol for women and a folk-heroine. This is still so, even seventy years after her death. However, to be fair to Jim Mollison, although largely forgotten today he was to become what many have described as the greatest long-distance solo flyer during a period now looked upon as the golden age of aviation. To say that he advanced himself by holding on to his wife's petticoat is to distort the truth.

The theory that their marriage was nothing more than a clever business arrangement can most certainly be discounted from Amy's point of view. The letters she subsequently wrote to her father, just before the marriage broke up, reveal a woman with a deep and genuine love for her husband. On his part it was probably no more than an infatuation. It went as deep as he was ever capable of going with a woman. His unstable upbringing might well have been a factor in his inability to sustain a relationship with the opposite sex. His parents' marriage had ended in divorce – quite a stigma in 1915 – owing to the fact that his father was a violent alcoholic. On one occasion, whilst still a toddler, his drunken father had threatened to throw him from the bedroom window of their first-floor apartment, and might have done so but for the timely intervention of the boy's nurse. When his mother finally remarried, he became the stepson of Charles Bullmore, which meant a name change, something he deeply resented.

Jim Mollison's subsequent career in the RAF as a nineteen-year-old pilot officer flying dangerous missions over rebel-held territory on the North West Frontier, did little to help. RAF aircrews operating in India in 1925 were expected to risk their lives in clapped-out, war-surplus machines without the protection of parachutes – small comfort when flying over rugged mountainous terrain which offered few opportunities for making a safe forced landing if an engine failed, and where the enemy beneath them rarely took prisoners. Alcohol then became for him a crutch and a means of subliminating what must have been deep-seated insecurities. It all added up to a man with problems.

When the news of their proposed marriage became known (they placed an announcement in *The Times* soon after they left Quaglino's) it caused a ripple of excitement throughout Fleet Street. On the following morning the corridors of the Grosvenor House Hotel were buzzing with journalists and photographers wishing to interview Mollison. Meanwhile, a delighted Amy was telephoning her parents in Bridlington and responding to calls from the *Daily Mail*, the newspaper that had done so much to promote her career. It was not long before the two aviators were described as 'the Flying Sweethearts' or 'the Air Lovers', and both were constantly being asked about their plans for flying together as a team.

One need look no further than the insensitive way Mollison broke off his engagement with Diana Wellesley to appreciate how he came to be regarded as the archetypal cad. The day after he had proposed to Amy, he even chose Quaglino's as the venue to break with Diana. One wonders just how she reacted as he tried to explain about Amy. No doubt he blamed Lady Cowley for the sudden switch. All credit must be given to Diana, for she announced to the press, with all the stoicism and stiff upper lip of her class, that she recognised the common interest that existed between the two aviators and wished them every happiness in the future. Diana eventually found someone of whom her stepmother did finally approve, and in the following year she married Daniel Dixon, a captain in the Grenadier Guards. She later became Lady Glentoran.

When Amy's parents realised that their daughter had known Jim Mollison for a total of less than twenty-four hours, they were perplexed by the speed of the engagement. It was in stark contrast to the prolonged courtship with

Hans Arregger. Jim's parents and relatives in the Glasgow area were anxious to meet his famous bride-to-be and the couple flew up to Renfrew in Amy's Puss Moth to meet his mother, Thomasina, and her husband, Charles Bullmore. No doubt the large, single, cinnamon diamond engagement ring on Amy's hand was studiously admired as she was introduced to the family. From subsequent correspondence it appears that Amy and Thomasina struck up an immediate and genuine friendship, although Amy's impressions of the rest of the family were that they were 'quite likeable, but rather dull'.

A month before their wedding Amy moved out of Vernon Court and began to enjoy the luxury of a suite of rooms at the Dorchester at a peppercorn rent. In the meantime, Jim suggested that they take a holiday together in Juan-les-Pins that June, but not before Amy had the difficult task of telling Jack Humphreys that he would no longer be required now that she was about to be married. If Amy needed any warning of the quality of her future life with Jim, then maybe she should have heeded his behaviour on that holiday. The heavy drinking and casual attitude towards debt were there to be seen, but she was either too infatuated with her suave, fun-loving fiancé to notice, or just chose to ignore it.

The more Jim and Amy were seen together in public the more the press began to speculate on a wedding date. There is evidence that Jim was beginning to have second thoughts about the marriage, because he admitted in his memoirs that he never had the same feelings for Amy that he had for Paula, his first love. He had met the seventeen-year-old whilst on a two-month spending spree on the Riviera after his discharge from the RAF in 1928. She was the attractive, red-haired daughter of a Russian emigré, one of many who fled after the revolution, and had fallen under the spell of the young Scot at a time when he was going through his small fortune of a gratuity as if there were no tomorrow. According to Mollison they became engaged and earnestly pledged that they would wait for each other and marry as soon as he found regular employment as a pilot. The marriage never materialised and he moved on to Australia.

Jim favoured a quiet wedding and was not prepared to give an invitation to any of his own relatives. Even in childhood his relationship with his mother had never been a happy one, and now in adulthood it was even more strained, largely because of his wild drinking habits, which to her were anathema. Whenever Thomasina visited him in London there was inevitably friction between them, because she would attempt to reform him. He obviously reminded her too much of his wayward father. She also resented his ties with the Mollison family who ostracised her after the divorce. Although she was largely the innocent party in the marriage break-up, she felt she was apportioned more than her fair share of the blame by her husband's parents.

It was against this background that Amy was influenced not to invite any of her own relatives to the wedding. In fact, she deliberately misled them by telling them not to believe the newspapers which reported that it was imminent, even when she knew otherwise. She may have felt some embarrassment after deciding that it was not to be a white wedding, which her mother would have expected. Amy's decision not to wear white would not have come as a shock to her father, who must have realised that her long-standing affair with Hans was an intimate one, but it might well have been a shock to the rest of the family.

On the evening of Thursday, 28 July 1932, the day before the wedding, Amy sent a telegram to her parents with a thinly disguised excuse that they had not been invited because she wished for a quiet ceremony owing to Jim's impending attempt on a North Atlantic solo flight. Several frantic phone calls then took place between parents and daughter during the late evening, before Will and Ciss Johnson decided that they and their two daughters, Mollie and Betty, would motor down to London in the early hours of the morning to be at the church on time. Driving through the night on wet, dangerous roads meant that by the time they arrived at Golders Green, where they intended to park their car and travel by tube and taxi to the church, there were only twenty minutes left before the service was due to begin in St George's Church, Hanover Square, Mayfair.

This high Anglican church had been a popular venue for society weddings ever since the poet Shelley had married the sixteen-year-old Harriet Westbrook there in 1811. It was also here that the Duke of Wellington acted as a witness and in the role of giving away a bride in the flurry of marriages which followed the Battle of Waterloo. On a less historical footing, *My Fair Lady*, based on George Bernard Shaw's *Pygmalion* features the rascally Alfred Doolittle being anxious to get to St George's, Hanover Square, 'on time'. Now before the high altar stood two people who were also making their mark on history.

The four members of the Johnson family arrived just as the service ended and whilst Amy and Jim, together with the best man, Lt Col Francis Shelmerdine (Brancker's successor) and Kathleen, Countess of Drogheda, were in the vestry signing the register. As the organ struck up the wedding march the couple walked slowly down the crimson-carpeted aisle smiling at well-wishers, but oblivious to the sad little group seated in one of the polished mahogany side pews at the back of the church.

Outside on the steps of the church the couple posed before a group of waiting photographers, who were exhorting them with: 'Come on Amy, give us a big smile' and 'Please look this way, Jim.' Amy obliged and beamed happiness. She stood looking stunning in a black coat-frock with silver fox-fur and matching black hat, neatly trimmed with an attractive eye-veil. The only concession to the all-black outfit were the white gloves. One cannot help but wonder if she was deliberately acting out the dream that she had whilst she was in love with Hans, where she saw herself being married in white gloves.

Amy's family did not have the opportunity to speak with her, even for a moment, before she and Jim were whisked off to their reception at the Grosvenor. However, the press were not slow to spot and recognise the four sad figures walking away from the church and to report what had happened in the early editions of the evening papers. The story must have disturbed Amy, for when it was brought to her attention, she had a telephone search made around the hotels which she knew her mother and father usually frequented when in London, but by this time the forlorn family were well on their way home to Bridlington.

One can gauge something of the hurt and humiliation that Amy's parents must have felt from the comment Will Johnson made when writing to their friend Mrs Glass a few days after the incident. He wrote:

> You will understand we cannot say much, but you, as a Mother, will have some idea of our feelings. Please do not be anxious about us, as we shall pull through. We can only conclude that both Amy and Jim have lost their heads. Meanwhile, they are in God's keeping and we know He will protect and watch over our dear little girl.

There was a further sad note when he wrote in his diary for the day after the wedding, 'Papers full of Amy's wedding. I don't feel very well.'

Amy's mother was equally wounded by not being invited to her daughter's wedding, and almost certainly blamed Jim Mollison. The fact that he had even refused to see some of his own relatives who had travelled down from Scotland purposely to attend the reception, seemed to confirm her view that the blame did not lie with Amy. If she did hold her daughter at fault, then she made light of it. One cannot help but be impressed by the gracious way in which she accepted Amy's excuse, in the following comment she made to a friend who had written to sympathise with her. 'Amy was very pleased that we had gone and *that* was all that counted to us. We only want to see her happy . . . She is in a different sphere now . . . if Amy lives a useful, happy life, we are satisfied and content.'

A toast to the newly married couple. The Mollisons' wedding reception was held in the Grosvenor House Hotel on 29 July 1932. (The Amy Johnson Estate)

The Mollisons spent only a weekend as honeymoon guests of Lady Muriel Bowden, the wife of the Birmingham industrialist Sir Harold Bowden, at Kelburn Castle, Largs, on the west coast of Scotland.* The reason for the brevity of the couple's stay was that Jim Mollison needed to prepare for his forthcoming Atlantic flight. He recalled that as they climbed the castle steps to its battlements and looked out across to the Western Isles and beyond, he could not ignore the sense of unease that he felt over the prospect of flying across 2000 miles of inhospitable ocean alone. He was supposed to be blissfully happy, but he was all too aware that not only had no one successfully made a non-stop crossing alone from east to west against the prevailing westerly winds, but no one had made it in a light aircraft. Only the two Americans, Charles Lindbergh and Amelia Earhart, had flown it alone, and that in the easier, west–east direction. In both cases they had used aircraft with much more powerful engines than the 120-horsepower installed in the aircraft Mollison intended to use.

* Lady Bowden had a lease on certain apartments in Kelburn Castle at the time.

Amy flew her de Havilland biplane from Stag Lane to Baldonnel on 16 August to join her husband, who had now decided to use the unrestricted stretch of beach at Portmarnock Strand, Dublin, as his runway. The Puss Moth had been fitted with two extra fuel tanks, both of which had been installed in the cabin, allowing just sufficient room for the pilot to squeeze in between them. The aircraft had enough fuel to make the 3176 miles from Dublin to New York, plus a little extra. It was literally a flying petrol tank. When writing about her husband some years later, Amy commented on his attitude to the hazards of taking off in overloaded aeroplanes.

> Jim Mollison never lacked the courage to take off on some flight with a heavier load of petrol than anyone had dared to take before. He would always say to the plane designers, 'You just put in the petrol I want, and I'll get the plane off the ground,' and he always did.

The Puss Moth's fuel load on the Atlantic crossing was equivalent to the weight of nine average-sized men, in a machine designed to carry a pilot and two passengers – a giant overload when one remembers that variable-pitch propellers and flaps for shortening a take-off run were innovations yet to be in production in the British aviation industry.

Amy's car arrived just before 10 a.m. on the 18th and descended to the beach to speed along to the Martello tower at the northern end of the strand, where a crowd had gathered around *The Hearts Content* to watch its departure. The name given to the silver-grey monoplane had a dual purpose. Not only was it meant to signify married bliss, it also coincided with a small village of the same name which was marked on his map as his intended landfall in Newfoundland.

The young couple embraced and whispered endearments as the newsreel cameras whirred before the cabin door was finally snapped shut. As Jim ruddered the aircraft round to face the 1½-mile stretch of sandy runway in front of him, he gave one last wave before opening the throttle. Within 1000 yards the tail was up and the wooden wings were lifting a ton and a quarter of aeroplane and its contents into a grey overcast sky. Amy stood gazing up as the tiny plane disappeared from view. She managed to conceal the anxiety she felt about the outcome of what she knew to be an extremely dangerous flight. She was all too aware that widowhood after less than a month of marriage was a distinct possibility.

Amy flew back to Stag Lane early the following morning to await the outcome of the flight, and was just about to have lunch in the Grosvenor when news came through that *The Hearts Content* had been sighted over Halifax, Nova Scotia. It was late afternoon before the telephone rang again in her apartment and this time she heard the voice of her husband phoning from St John in New Brunswick, Canada. He had been in the air for thirty hours before landing in a large, rolling meadow in a sparsely populated farming settlement at Pennfield Ridge. He had flown some 2650 miles and there were just ten gallons of fuel left in the tanks.

Not only was Mollison's flight the first non-stop solo across the Atlantic from east to west, but it was also the fastest. Moreover, it was the first crossing in a light aeroplane and the longest non-stop flight in one. It is not without justification that his crossing of the North Atlantic in 1932 has been described as one of the greatest solo flights in the history of aviation.

Jim's success only made Amy all the more keen to add to her own laurels as an aviatrix. Although she was still considered a national heroine, she had, since her Australian and Tokyo flights, faded somewhat from the public's eye. She telephoned Jim to say that she would like to attempt a similar solo crossing and match Amelia Earhart's performance earlier in the year. However, as soon as Jim heard of her plans, he did his best to dissuade her. Contrary to popular belief, he was not the complete male chauvinist that some portrayed him to be, and on this occasion he was concerned purely for her safety. On her subsequent attempt on the Cape record he did all he could to encourage her, even when he knew that his own record was likely to fall to her.

Amy accepted her husband's advice because she realised that she lacked the ability to fly blind by the use of instruments alone, a skill that would be essential whilst flying during darkness over such a large stretch of ocean. So determined was she to remedy this deficiency that she took the opportunity, during the time that Jim was away, to enrol with the Air Service Training School at Hamble, near Southampton, for a course of blind-flying instruction.

For the next four weeks she took sporadic tuition in one of the AST's black and silver Avro Cadets, a small biplane used for *ab initio* instruction in blind-flying and aerobatics. In this type of machine it was customary for the pupil to occupy the rear of the two open cockpits. Once in the air a folding canopy could be pulled over the rear cockpit, leaving the pupil completely blind apart from the illumination of the instruments. Corrections of any error in flying attitude or navigation by the pupil could then be conveyed over the speaking-tube system by the instructor. Amy was able to complete almost ten valuable hours of this type of instruction.

Amy's course at Hamble was interrupted when she flew to Cherbourg to meet her husband on his return from Canada in the liner *Empress of Britain*. It had been his intention to fly back in *The Hearts Content* and so achieve the distinction of being the first person to make a double crossing of the North Atlantic alone. He had in fact set out to do so, but by the time he reached Cape Breton Island, where he encountered violent thunderstorms, his nerve gave out.

He was suffering from nervous exhaustion, and the doctor who was called to treat him sent a cablegram to Amy in London which read, 'I strongly urge that he should not attempt the return trip until such time as he recovers complete control of his nerves. I think you should insist on this.' Amy made several transatlantic phone calls imploring Jim to give up, but it was only when the weather reports confirmed that conditions over the Atlantic were likely to get worse that he finally threw in the towel. He is quoted as saying, 'I felt as if some sickening, intangible weight had been lifted off my head.'

The Canadian Pacific liner was due in at its first port of call at Cherbourg before proceeding to Southampton, but such was Amy's excitement at being reunited with her husband that she chartered a private aircraft to fly her to meet him at the French port. Once she arrived there, she found that owing to rough seas the liner was holding off and had not yet docked. Not to be put off, she hired a tender and braved the elements to go out and board the ship outside the harbour.

These are hardly the actions of a woman who married purely for business reasons. There is no doubt that she was deeply in love with her husband, and he, in his own strange way, with her. Hans was now just a distant memory from her youth, and the emotional wounds that he had inflicted had already healed. The problem was that Amy's and Jim's views of marriage and what was required from it did not really coincide. In spite of her independent adventurous spirit, Amy was, as we have already seen from her obsession with her former lover, a romantic at heart. She was vulnerable where men were concerned. On her husband's part, agreements, appointments, decisions and even marriages were transient, merely arrangements which could be broken if and when it suited him.

Amy soon discovered that married life with Jim was to become very much a 'goldfish bowl' existence. It was a life she hated, but it did not seem to bother him in the least. There was a constant stream of visitors to their apartment and they were rarely if ever left alone. They needed time together if they were to get to know each other and this was denied them right from the start. It was a major stumbling block in the marriage and an obvious recipe for marital disharmony. James Mollison Steven, a first cousin of the aviator, threw some light on this aspect of the marriage when he made the following comment, 'James was very much in the hands of the press and they ruined him. They lauded him so much, he just thought that he could sit back in life and entertain his friends.'

Jim and Amy Mollison visit the 'Silver Lady's Night Cafe' in London after an evening at the theatre in 1932. (Hulton Getty Picture Collection Ltd)

One can grasp the scale of the couple's popularity at this time from the fact that immediately after Jim Mollison's transatlantic flight, he was asked, much as Amy had been before him, to make a radio broadcast to the nation. More surprisingly for those days, he was also asked to appear before a new medium of broadcasting–the television camera. Although there was no regular television service in 1932, the BBC was making experimental late-night transmissions in collaboration with the Baird Company.

That same evening, Jim and Amy attended a variety show at the London Palladium, where they delighted theatre-goers by appearing together in the royal box. They were probably at the height of their appeal with an adulating public, since they were unique as a husband-and-wife team in pioneering aviation. One must not forget that their combined achievements were ultimately to surpass even those of the Lindberghs, for although Anne Morrow Lindbergh was a qualified pilot who flew with her husband on several geographical surveying flights, she was never to achieve independent recognition as an aviatrix in her own right. Her main interest lay more in the realm of being a serious writer and a home-maker for her children.

Bill Courtenay believed that if Amy had been able to have children, then things might have been different in her marriage. Although it was never openly admitted that she had undergone a hysterectomy, it is significant that, in spite of expressing a longing to have children, she was never to have any of her own. The couple's close friend, Charles Kingsford Smith, had around this time become a father, and the Mollisons had sent him and his wife a telegram of congratulation. The wise-cracking Australian immediately wired them back with a short message which simply read, 'Jim and Amy Mollison, Grosvenor House, London. Luke 10, verse 37. Smithy.' As they turned the pages of a bible to look up the quotation, it read, 'Go, and do thou likewise.'

Sadly, the line, 'A boy for you, a girl for me', in the popular ballad 'Tea for Two', was never to become a reality for Amy.

The record for the fastest time to the Cape was held by Jim Mollison, and the record for flying there and back was held by Captain C.D. Barnard and the Duchess of Bedford, and Amy was determined to break both of them. This time there were no financial backers and she had to spend £2000 of her own money to fund the adventure. De Havilland sold her a brand new Puss Moth with an uprated engine, one that would give her a 5 m.p.h. speed advantage over the similar machine that Jim had used earlier in the year. This time she dropped the *Jason* appellation, having used it on four consecutive aircraft, and christened the silver-grey monoplane *The Desert Cloud*. The only concession to the memory of her earliest machine was that the aircraft's G-ACAB registration letters were painted in a stark, bottle-green colour.

Whilst it cannot be denied that Jim Mollison gave Amy the benefit of his experience in using the shorter, west coast route of Africa to the Cape, he still harboured a feeling that now she was married she should accept the conventional role of a wife. This comes across in, *Playboy of the Air*, where he intimated, probably with a fair degree of tongue in cheek, that she had deserted him to fly off to the ends of the earth within four months of 'taking him to the altar'.

Maybe this was no more than how Scottish men traditionally viewed their wives in the early thirties, or perhaps he failed to appreciate fully that this was no ordinary woman he had married. However, there was little he could do about it for she was financially independent and he took the view that if anyone was going to eclipse his record, then maybe it was better that it was she who did it. He supplied her with the maps which he had used, gave her advice on the primitive landing grounds she would encounter, and arranged for the long-range fuel tanks that he had used on his own flight to the Cape to be fitted to her machine.

Bill Courtenay and Jim Mollison flew down to Lympne in *The Hearts Content* on the afternoon of Sunday, 13 November to be ready for Amy's departure on the following day. Amy and Jim, both immaculately dressed in long black leather coats trimmed with astrakhan to keep out the cold, made their last-minute inspection of her machine before driving away to snatch a few hours' sleep at a local hotel. Not surprisingly, the inability to sleep just before a long flight appears to have been an occupational hazard with most of the pioneering, long-distance aviators. Charles Lindbergh found it virtually impossible to sleep just before his epic New York–Paris flight, and Amy had found similar problems at the time of her Australia flight. Only Jim Mollison seems to have overcome the tension by the generous use of alcohol.

Just before daybreak the silence of the Kent country-side was broken as

the engine of *The Desert Cloud* burst into life. Slowly, the young woman who was to dominate the headlines for the next four days taxied her plane out of the hangar towards a small waiting crowd. Before her lay a journey which would take her 6700 miles across mountain ranges, deserts and jungles. The stakes were high, since she carried neither parachute nor radio, and a forced landing in any of these areas could easily prove fatal. Even if she did manage to put down successfully in the Sahara

Amy at Lympne with her DH Puss Moth, The Desert Cloud *(G-ACAB), just prior to her departure on her record-breaking flight to the Cape and back by the west route in November/December 1932.* (Author's Collection)

after an engine failure, there was no guarantee that she would be found, since the radio facilities for rescue were inadequate.

Amy intended to make her destination in five long hops: from Lympne to Oran in North Africa; across 1900 miles of the Sahara desert to Gao in French West Africa; down to Douala in the Cameroons; southwards to Mossamedes on the coast of Angola; and finally to Cape Town. By taking the west-coast route, rather than what was known as 'the all-red route' used by Imperial Airways, which ran southwards from Cairo to the Cape, via Khartoum, Kisumu and Johannesburg, she would cut approximately 600 miles off the flight distance.

At 6.37 a.m. the faint outline of Amy's tiny monoplane was barely visible against the twinkling red boundary lights of the aerodrome, as it took off into the darkness and disappeared from view. Jim was already above the aerodrome circling and waiting to accompany her across the Channel, but within ten minutes of her departure, he was back on the ground reporting that he had been unable to sight her aircraft in the murky conditions. All that he and Bill Courtenay could do now was to fly back to London and wait.

The first messages from Reuters and the Press Association reported that Amy had put down at Barcelona with a suspect fuel gauge, but it later transpired that pressure in the upper wing tanks had caused fuel to overflow from the breather pipes. It was not a serious problem and she was only on the ground for a matter of minutes before proceeding southwards to cross the Mediterranean and land at Oran by 7.30 p.m.

After a four-hour delay on the ground, refuelling and satisfying the French authorities about her flying permits and documentation, came the long flight in darkness over the high barrier of the Atlas mountains and on across the Sahara. Flying this stretch of the desert by night had the advantage of avoiding the worst of the heat, but the combination of sleep-deprived eyes and the dance of the luminous instrument needles had a hypnotic effect upon a tired pilot. It seems that for this part of the journey, when the air conditions were reasonably smooth, Amy felt the benefit of her blind-flying tuition, but when she encountered turbulent tropical storms and severe buffeting later on in the flight, she had her doubts. Jim believed that Amy had the skill to break his record, but he wondered if she had the stamina to do so.

Amy breathed a sigh of relief as she circled the landing ground at Gao on the banks of the river Niger on the following afternoon, believing that the worst part of her journey – 1000 miles of desert heat – was now behind her. She touched down in a swirl of dust on a large arid expanse of what passed for a primitive aerodrome, its perimeters marked out with whitewashed boulders to distinguish it from the rest of the surrounding scrubland. Her intention was to snatch a few hours' sleep whilst her machine was being refuelled, and make a night flight to Douala, a distance of 1200 miles. As it transpired, she did not leave until the following morning.

There was some dispute about events at this point. Amy maintained that when she left Gao at 8.40 p.m. and had been in the air for thirty minutes, she found that her fuel tanks, which had a maximum capacity of 119 gallons, were insufficiently filled. However, the Shell Company's agent at Gao disputed her story and was adamant that they had filled all but a quarter of one of her interior tanks, making a total of 100 gallons supplied. If true, it was certainly enough for the next stage of her flight. One is forced to ask why she did not check the fuel gauges before taking off. It might well have been that she was so tired after the Sahara crossing that once in the air again she decided to turn back and spend the night at Gao.

The physical disturbance caused by so many continuous hours in the air, in a small aeroplane flying long distances is well illustrated by Alex Henshaw, a man well qualified to comment by virtue of his own record-breaking flight to the Cape and back in 1939. In his book, *The Flight of the Mew Gull*, he said he found that, back on the ground after almost eleven hours of flying, he became very disoriented. Whilst eating a meal he had difficulty in dispelling the vertiginous after-effects of so many hours in the air. The combined effect of the continual drumming in his ears, and the illusion that the table at which he was sitting appeared to be in constant motion, was not a pleasant experience. Maybe this was how Amy felt on the night she took off from Gao; if so, who could blame her for turning back.

She arrived over the port of Douala late on the afternoon of the 16th to find a small airfield set in a clearing of the jungle and surrounded by palm trees. One of the things she had to take into account whilst flying in equatorial latitudes was the rapidity with which a tropical sun could sink below a tree-lined horizon. Timing was of the essence. To arrive late and then attempt a night landing without adequate ground lighting facilities could prove fatal.

Within less than two hours, she was refuelled and back in the air for a fourteen-hour flight through the night to Mossamedes in Angola. It was to be the worst part of the whole journey. She ran into tropical storms which buffeted her plane mercilessly around the sky, as heavy rain bounced off its wings. It would have been relatively easy for her to fly around such storms in daylight, but not so in the darkness. Visibility was so poor that whenever she attempted to follow the outline of the breakers on the coastline, she would find her port wing coming perilously close to cliffs, which seemed suddenly to jut out from nowhere. Under such conditions she was forced to gain height and fly blind by instruments, not as easy as she had found it over the Sahara.

At one point she discovered that she was off course by almost 100 miles, and what was more, she was out over the ocean. As soon as she picked up the coastline again, and not long after she passed the airfield at Benguela, some 200 miles north of Mossamedes, she noticed that her oil-pressure gauge was reading alarmingly low. Wisely, she made a 180-degree turn and flew back to investigate the problem. The landing turned out to be a tricky one, because the airfield had been struck by one of the rainstorms she had encountered and had become literally a sea of mud. To make matters worse the heavy rain continued throughout the day, and as a result it took nearly nine hours to rectify what was simply a matter of a clogged filter.

At this point Amy was on the verge of giving up. She cabled her husband to say that although she was only 1440 miles from Cape Town, she might well have to abandon the whole flight. This message was picked up by sceptics, some of whom were only too ready to say that it was foolish for a woman to think that she could accomplish what had fatigued the best of male pilots. Meanwhile, the phones were ringing continually in the newspaper offices around the world, with one question, 'Where is she now?'

By 4.20 p.m. Amy had left Benguela for Mossamedes, where she rested and refuelled, ready for the last leg of the journey. Eating was always a problem on these long, arduous flights. Local food was usually avoided for reasons of hygiene, and Amy had so far survived on coffee, together with sandwiches, almonds, raisins and barley sugars. Whether she relented and ate local food at her last two stops we are not told, but it was not long after she departed from Mossamedes in the early hours of the morning that she suddenly felt faint. Fortunately, she carried items for such an emergency and was able to pull herself round with the assistance of smelling-salts. Liberal applications of eau-de-cologne to her face kept her awake for the final eleven hours to the Wingfield aerodrome at Cape Town, and knowing that she would be greeted by well-wishers, she even managed to powder her nose before landing.

Just before 3.30 p.m. *The Desert Cloud* was seen approaching from the north and following the sweep of Table Bay, before coming in to land. Her flight had taken four days and seven hours, and she had beaten her husband's record by almost ten and a half hours. The crowd had been forced to wait behind newly erected barbed wire fences, but even before the engine of her Puss Moth was switched off, the fences were broken down and a surging mass of hundreds of people swarmed towards the plane to greet her.

The press back home acclaimed her as 'the world's greatest woman flyer' and as 'the Queen of the Air', and there were world-wide tributes to her success, with over three hundred cables of congratulation waiting for her when she arrived in Cape Town, including those from her family and one from Amelia Earhart. Jim Mollison was equally ungrudging in his praise, for she had shown that she could compete with the best of the men and equal their physical stamina. She had managed to get no more than five hours'

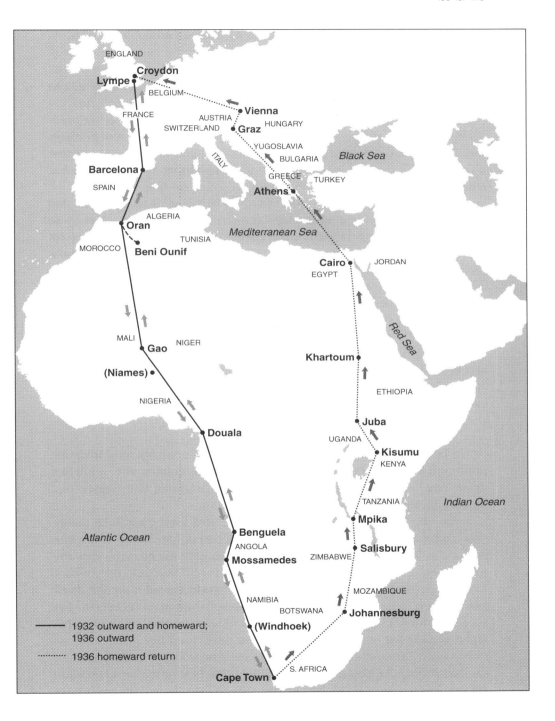

———	1932 outward and homeward; 1936 outward
··············	1936 homeward return

211

sleep during the whole of the flight, and yet stepped down from her plane looking remarkably fresh.

One is bound to ask at this point exactly what Amy hoped to achieve by this flight. There were of course the usual platitudes about establishing links with the 'far distant Empire', and of making the public 'air-minded', but the one thing she wanted above all else was to be taken seriously in her quest for a career in aviation. As a woman seeking such a career, she was probably twenty years ahead of her time. To follow the example of her friends, Pauline Gower and Dorothy Spicer by merely giving pleasure flights around her own country, did not appeal to her at all.

Obviously there were monetary rewards to be gained from her flight, and she left this task to her manager, Bill Courtenay. The difficulty for him lay in the fact that newspapers such as the *Daily Mail* and others on Fleet Street were no longer prepared to offer large sums of money for rights to her story. Such flights were becoming increasingly commonplace, with flyers such as Jean Batten now coming on the scene. Moreover, Amy had not fulfilled her contract for the aerial tour of Britain after she had returned from Australia in 1930. The best that Courtenay could do was to take what reduced sums there were on offer and rely on enhancing her finances by royalties from advertising. Amy was shrewd enough to realise that even this income could not last, since the days of what were derisively termed 'the stunt-flyers', were fast coming to an end.

Amy decided to wait for the next full moon before making an attempt on the record for the return flight, and as a consequence she spent the remaining weeks enjoying the sunshine and blue skies as the guest of a wealthy local family whom she had met in London.

It was not quite the idyll that she would have liked, for she admitted that she was homesick and missing Jim. There are tell-tale signs, reminiscent of her time with Hans, that she felt that her love and affection were not being reciprocated as fully as she would have wished. She was overheard talking to her husband on the telephone, soon after landing, about when she should make her return flight, saying, 'Now I have lost the moon and have no plans . . . Do tell me what to do, Jim.' One speculates whether she was seeking genuine advice or merely trying to make her husband feel he was being missed.

It was a bronzed and healthy-looking Amy who set out from the Cape on 11 December, for Croydon. She was fairly confident of breaking the existing record on the return flight, for she had only to beat the rather leisurely ten days set by Captain Barnard and the Duchess of Bedford. Even so, her flight was not an easy one, because on the initial stretch she had to fight her way through severe weather between Mossamedes and Douala. Whilst trying to make Oran non-stop across the Sahara from Gao, she encountered blinding snowstorms over the Atlas Mountains. These forced her to turn back and land at Beni Ounif, a French Foreign Legion outpost on the fringe of the desert. Much to her dismay she found that the emergency landing ground was being battered by gales, and the delay she experienced here forced her to give up all hope of beating her own record. She eventually arrived at Croydon in bright sunshine and was given a rapturous welcome. It was seven days and seven hours since she had left Cape Town.

Jim Mollison ran out from the crowd towards the travel-stained silver Puss Moth as Amy taxied in. He was every bit the proud husband. They kissed and embraced affectionately before she slid her arm in his, and as they walked together towards the cheering crowd she was overheard to say, 'They should be cheering you, Jim. They don't know how easy your advice made it.' On the reception platform waiting to greet her were her parents, Jim's mother, Lady Drogheda, Sir Malcolm and Lady Campbell, Bill Courtenay and the usual Air Ministry officials, one of whom conveyed the king and queen's warmest congratulations.

Once more Amy was back in Fleet Street's headlines as the press recognised her gutsy performance. She had taken on the men, principally her husband, and beaten them at their own game. She had become the first woman to solo the trans-Sahara route in both directions, and into the bargain had gained two new records on a 13,000-mile flight. The crowd were ecstatic, and chanted 'Amy! Amy!' as she and Jim smiled and waved from the back of their chauffeur-driven car, as it sped off towards the West End of London, leading a procession of cars in its wake.

Again, fortune had favoured the brave.

(Chas. R. Pickard & Son)

Chapter fourteen

TICKER-TAPE ON BROADWAY

he Mollisons needed little encouragement to make the most of their
winter sports holiday in Switzerland at the end of 1932. On the way
over they stopped off for a day in Paris, where Amy was able to shop
on the Rue Cambon and to visit Gabrielle 'Coco' Chanel's boutique. The
Parisian designer was the acknowledged queen of *haute couture*, attracting
the custom of some of the wealthiest and most fashionable women on both
sides of the Atlantic, including Hollywood stars. Amy delighted in the
simplicity of the couturier's creations, particularly her current fashion of the
garçonne look, a vogue using black, Amy's current favourite colour. Chanel
was highly flattered by a visit from someone whom France regarded as the
aviatrice extraordinaire, and offered Amy the choice of any of her evening
gowns as a personal gift.

Christmas at the sumptuous Palace Hotel in St Moritz saw the Mollisons
rubbing shoulders with some of society's *glitterati*, amongst them Gertrude
Lawrence, Cicely Courteneidge and her husband Jack Hulbert. Hollywood
stars included funny-man, Harold Lloyd, the beautiful Ann Dvorak and the
red-haired Clara Bow, a woman on the cusp of her career, having fallen
victim to the transition from silent fims to 'talkies'. It was just as well that the
Mollisons' marriage was at the peak of its harmony at this time, for one can
only imagine what temptation the latter lady might have posed for the Scot
under anything less than the vigilant eye of a wife. She was Hollywood's
original 'It Girl' – 'it' being a synonym for sex-appeal – and the first in a long
line of sex-goddesses.

That the Mollisons were on an equal footing with celebrities of stage and screen cannot be overstated, for they too were stars in their own right and the public recognised them as such. Apart from newspaper articles detailing the minutiae of their lives, they featured in the weekly *Pathé* newsreels shown at every local cinema and their pictures appeared regularly in widely read society magazines such as the *Tatler* and the *Illustrated London News*. At the time of their holiday much publicity was given to the fact that Amy was soon to collect the Segrave Trophy, a prize awarded annually to the person who had contributed the most to Britain's exploits on land, sea or air. The 1932 award went to Amy, whilst in the following year it went to Jim.

It was whilst Amy was at St Moritz that she risked her neck on the Cresta Run disguised as a man. Originally, women had been allowed to compete in the Ladies' Grand National event in the head-first prone position, but in 1929 a ban was imposed on them doing so. The decision was taken purely on medical grounds because it was felt that for a woman to lie face downward on a toboggan, with the inevitable pressure on their chests at speeds of up to 80 m.p.h. over anything but a smooth surface, was a possible cause of breast cancer. However, in spite of the ban, many women, Amy among them, were reported as having dodged the scrutiny of the management and taken part clandestinely. Other observers had seen women riding the Cresta Run at night. Bob Ennis, an American who used to test-drive some of the world's fastest cars, once said, 'The illusion of speed on the Cresta is greater than driving a Ferrari at 200 m.p.h.'[16]

Jim and Amy at St Moritz in January 1933. (The Amy Johnson Estate)

Whilst Amy revelled in the sunshine and pure air of the mountains, and would have been prepared to extend the holiday, Jim, never particularly fond of sports, was becoming daily more restless. He was anxious not to rest on his laurels, but to return to London, where he was preparing for his next major flight. Four prizes now lay tantalisingly within his reach: to become the first person to solo the South Atlantic westwards; thereby becoming the first person to have soloed both north and south stretches of that ocean; to be the first person to fly from the UK to South America; and to beat the time of four and a half days to reach Brazil, a record set by a French crew flying from Paris in a Couzinet tri-motor. Not content with his own plans, he had also agreed with Amy that they should place an immediate order with de Havilland for a twin-engined aircraft, in which they would make a joint attempt on the world's long-distance record later in the year.

As he prepared his Puss Moth machine for the South Atlantic crossing during January 1933, one piece of news must have disturbed both him and Amy. Early in January, Bert Hinkler had set out in a similar aircraft in an attempt to smash Charles Scott's England–Australia record. Rather foolishly, he had commenced his flight in atrocious weather conditions to cross Europe via the Alps, with Brindisi as his first planned stop. He was never seen alive again. After several days' waiting for news of the missing pilot, an aerial search party began to scour the snow-covered mountains and valleys but without success.

It became clear from the wreckage, when it was found four months later at Pratomagno in the Appenines, that Hinkler's Puss Moth had suffered a structural failure whilst in flight. The port mainplane was discovered several hundred yards away from the rest of the aircraft, and it appeared that Hinkler, although badly injured, had survived the crash. Evidently, the little Australian had made a brave but vain attempt to crawl away from the wreckage and had died from exposure in the extreme cold of the desolate mountainside. It was not the first of a disturbing number of crashes where a Puss Moth had shed a wing whilst in flight, and in spite of modifications which the de Havilland factory had begun to implement, the crashes were to continue. Perhaps it was just as well that the wreckage of Hinkler's aircraft was not found until three months after Jim Mollison had departed for South America, but it is significant that neither he nor Amy ever used the Puss Moth again for long-distance flights after the cause of Hinkler's death was discovered.

The Puss Moth had been known for its excellent handling qualities, and became a firm favourite with light aeroplane owners throughout the world in the early 1930s. Unfortunately, its reputation became tarnished when instances of accidents similar to Hinkler's became known. One of the earliest cases of such a failure occurred in Western Australia in October 1930, but at the time it was considered to be an isolated incident. However, when the well-known pilot, Lt Cdr Glen Kidston, and his passenger were killed in South Africa in May 1931, the manufacturer was forced to look closer into the problem.

At the Board of Inquiry investigation into Kidston's death, allegations were made by an inspector of air accidents to the South African Air Force of an 'inherent weakness' in the aircraft's design. The charges were vehemently denied by the de Havilland Company at the time, but by the end of 1933 there had been nine known instances of wing failure. As a result, the Royal Aircraft Establishment carried out several investigations during the period 1931–4, in which the torsional stiffness of the Puss Moth's wing was compared with that of other contemporary aircraft. The outcome was that the company was forced to recommend modifications to the rear spar and tank bay areas of the wings, both in production and retrospectively to existing aircraft.

In spite of the modifications, the Puss Moth, was never fully able to shake off its reputation for being prone to wing failure. The last known case was reported in June 1937, when an owner and two passengers were killed whilst on a flight from Toronto to Detroit. It appeared that failure was most likely to occur under the combined conditions of high speed and turbulence. These must have been exactly the kind of conditions that Amy experienced during her night flights over West Africa, without her being aware of a suspected design fault in the machine she was flying. Small wonder that in a letter to her parents she once wrote, '. . . however irreligious and careless I may seem nowadays to the outward eye, yet I know perfectly well it was my great trust in God which brought me safely through.'

Early on the morning of Monday, 6 February, Amy's *Desert Cloud* dipped in salute as she waved farewell to her husband and his aircraft somewhere near Le Touquet. Jim Mollison went on to achieve all four of his targets by crossing the South Atlantic and touching down at Port Natal in Brazil. It was three days and ten hours since he had left Lympne and he had the satisfaction of knowing that he had made the first flight between England and South America. Moreover, he had beaten the Frenchmen's record by more than a day.

Whilst Jim was being fêted in Rio, with an invitation to meet Brazil's provisional president, Bill Courtenay was accompanying Amy in her Mercedes on a publicity tour to the north of England. She had been invited to her native Yorkshire for an official welcome by the City of Leeds. The programme included a luncheon given in her honour by Lewis's store, where her aircraft was currently on display to the public. One imagines that there was possibly just the trace of a smile on Amy's face as she and Bill Courtenay entered the store. Only six years previously she had worked for the same firm as an insignificant learner in the Silks Department of their Peter Jones subsidiary in London; now she was being given VIP treatment.

Those who were keen to recruit Amy to the ranks of nascent feminism would find small comfort from the way in which she never objected to being referred to as Amy Mollison, as in the case of her Leeds visit. She was quite unlike her American counterpart, Amelia Earhart, who was rarely if ever publicised other than by her maiden name. By contrast Amy was content to let her achievements speak for themselves; she never felt that her reputation was threatened by her marriage.

Amy and Bill Courtenay motored over to Blackpool on the evening of their Leeds visit, to call on her sister Mollie and her fiancé, Trevor Jones. Trevor was the Senior Assistant Solicitor to the City of Hull when he first met Amy's younger sister, and had recently been appointed as the Deputy Town Clerk in Lancashire's favourite seaside town.

Bill Courtenay recalled a rather amusing account of the interview in which the young Welshman was selected for the job. Evidently the Mayor of Blackpool was on the interviewing board, and he must have had some slight misgivings about the suitability of the candidate sitting in front of him. After a while he came to the conclusion that whilst the young man was quite capable, he would have preferred someone with the stability of a marriage behind him.

'Well,' said the Mayor, 'although you come from Yorkshire, we have some fine girls here in Lancashire and no doubt Lancashire can provide you with a wife.'

'But,' said Trevor, 'I already have a girl in Hull.'

'It's not Amy Johnson by any chance, is it?' said the Mayor.

'No, but it's her sister,' came the reply.

Needless to say, he got the job![17]

After spending the weekend with her sister in Blackpool, Amy motored back to London to be in time to catch the liner *Highland Brigade* from Southampton to Madeira. She had made a pact with Jim before he flew off that he would break his homeward journey by sea from Brazil and she would meet him at Funchal.

These were the halcyon days of the Mollisons' marriage.

Edward Henry Hillman had started out in life as a semi-literate farmer's boy, but became an enterprising coach driver who eventually scraped up enough money to buy his own coach. By the late 1920s he was the owner of a fleet of private coaches plying on regular scheduled routes into London from the eastern suburbs. He had managed to break the monopoly of the London Passenger Transport Board's bus service, and subsequently a sizeable proportion of the city's daily commuters forsook the ubiquitous red buses in favour of his cheaper service. The Hillman coaches also provided a faster service, made possible by having fewer stopping places along the route.

By 1932 Hillman's fleet of coaches was the second largest in the UK, and he was the largest private owner of motor coaches in the world. However, his success was short-lived, for new legislation placed heavy restrictions upon his service, to such an extent that he was eventually forced to close down his entire bus operation. The closure was a big disappointment to him, but was softened by very generous financial compensation which, together with the sale of his fleet of vehicles, amounted to £1 million – a sizeable sum of money for those times.

Hillman was not a man to give up easily, and he was far-sighted enough to turn his attention to air transport when he could see the axe about to fall on his coach services. The result was that he changed the name of his company to Hillman's Saloon Coaches & Airways in November 1931, and began to operate an air-taxi service to Clacton and Ramsgate out of Maylands airfield near Romford. The aircraft he used were three Puss Moths, which only carried two passengers, and three of the more economical Fox Moths, which carried four.

Hillman was one of the first to have the vision to provide cheap air travel for the masses, much as Freddie Laker was to do in the 1970s. To fulfil his dream of running a daily service to Paris, he needed bigger and more economical aircraft, and in the summer of 1932 he approached de Havilland to see if they could help him. Fortuitously, the company was laying down the design for a twin-engined military aircraft for the Iraqi Air Force, which was easily adapted into a light airliner with the capacity to carry six passengers. Hillman, who knew absolutely nothing about aircraft, immediately ordered four straight off the drawing-board.

The prototype of the legendary de Havilland Dragon made its first flight at

Stag Lane on 12 December, and was delivered to the Romford airfield eight days later, where the blue-and-white-liveried biplane was christened *Maylands* by Amy. Obviously her business manager had kept his ear to the ground and struck some kind of a deal with Hillman for her to appear along with Jim, just two days after she had returned from her Cape flight. It was good publicity for both Hillman Airways and the Mollisons.

Another factor prompting the Mollisons' attendance was that the Dragon was the aircraft that they were contemplating using in their attempt on the world's long-distance record. Moreover, Amy saw in Edward Hillman the man who might be able to further her ambition to achieve a serious career in aviation, first as one of his airline pilots but ultimately with a seat on the board of directors of his company.

At the time when Amy and Jim were considering making an attempt on the world's long-distance record, it stood at 5341 miles, set by two young RAF officers when they flew from Cranwell to Walvis Bay in South Africa non-stop. Record breaking was now becoming relatively commonplace and the Mollisons knew that if they were to achieve any distinction at all, then they must make a flight that would be considered truly spectacular. Therefore their aim was to fly from the UK to New York, the first to do so, and then to fly the 5994 miles from New York to Baghdad non-stop for the long-distance record. With a subsequent return journey to the UK of approximately 2500 miles, they hoped that the whole flight would constitute a grand total of almost 12,000 miles. Comprising as it did, a double crossing of the North Atlantic, it was sure to attract the kind of publicity that the Mollisons needed. Bill Courtenay was enthusiastic and determined that they should receive the maximum press coverage. To this end, he managed to tempt the *Daily Mail* into agreeing to pay a substantial fee for the exclusive rights to their story if the flight was successful.

The Mollisons went ahead on the strength of what they had seen at Maylands, and ordered a DH Dragon to be specially modified at Stag Lane for its long-range task. The price of the aircraft to Hillman had been £2800, plus £105 for six comfortable passenger seats. From documentation in the archives of the RAF Museum at Hendon, we know that the Mollisons paid exactly half this sum, and although the aircraft was registered in Amy' name, it appears that they split the cost equally between them. Obviously, the de Havilland Company were looking for a pay-back in sales as a result of publicity from a successful outcome of the couple's flight.

It was whilst they were waiting for delivery of the machine that Amy became the target for some scathing public criticism for one of her earlier flights. It concerned the ill-advised flight that she had made in January 1931, when she crash-landed in the snow-covered potato field near Warsaw. She had recently put her name to an article for the *Sunday Dispatch*, detailing the incident in what was said to be a lurid and exaggerated manner. She had evidently described how she had at first been surrounded by a dozen or so peasants, one of whom was 'a big, dirty, bearded, evil-looking man with greedy eyes'. She alleged that he had not only demanded money, but had attempted to drag her from a sleigh into a hut. Whether she intended it or not, her account gave the impression that she had been the victim of a near rape. When the article was brought to the attention of the Polish Aero Club, they set up an investigation, and as a result of their findings her allegations were seriously challenged and a complaint made to the Royal Aero Club.

The outcome was that Amy was forced to say that the facts had been grossly distorted by the newspaper without her knowledge, and in spite of a letter of apology to the Polish Aero Club in which she withdrew the allegations, she was heavily censured by the Royal Aero Club. Its secretary, the irascible Commander Harold Perrin, informed Amy in a none-too-courteous manner that she could not evade responsibility for articles which were signed in her name. Jim sprang to his wife's defence by telling reporters that his wife had been 'made nervous' by the bearded man's action and stressed that she had, after 'the incident', experienced the utmost hospitality from the Polish people. The whole story was circulated in the world's press and it was eagerly seized upon and used against the Mollisons on later occasions.

The Mollisons' new Dragon, now given the name *Seafarer*, created quite a stir of excitement when it was flown on its initial trial flight in April to the British Hospitals Air Pageant at Boughton, near Northampton, for its first public outing. The twin-engined biplane had been drastically modified by fitting three massive aluminium fuel tanks into the space normally occupied by its passenger seats. Each of the 200-gallon tanks was suspended from the top longeron joints of the fuselage by steel cables, and separated from each other by sorbo rubber pads. They left a gap of only eighteen inches at their lowest point for Amy and Jim to crawl under when making their way to the cockpit. It was not exactly an easy escape route in an emergency. Small wonder that Jim Mollison had nicknamed the all-black machine 'the flying coffin'; he was fond of adding that 'it only needed brass handles to make it look like the real thing'.

At the end of May they took the *Seafarer* on a nine-hour cross-country proving flight in order to establish the aircraft's most economical cruising speed. It also gave them the opportunity to observe the result of switching fuel supplies from tank to tank during the flight; and to see what effect, if any, the surging of the fuel had on the supply to the carburettors. It was estimated that they would be burning approximately 12 gallons per hour at their optimum cruising speed of just over 100 m.p.h. In the cockpit, a wooden tip-up bench seat with canvas backs had been substituted for the single pilot's seat used in a standard Dragon, so that pilot and co-pilot could sit side by side. Inevitably, it meant that during the long transatlantic flight they were sitting for long periods in a cramped position, which could only be relieved when they took turns at resting on a canvas camp bed beside the front fuel tank.

Just before their Atlantic crossing *The Aeroplane* and *Flight* aviation journals put out disparaging comments about the proposed flight. Their editorial staff not only questioned the value of the flight, but also doubted the couple's chances of even surviving the ocean crossing. Amy was already smarting under the Royal Aero Club's criticism of the Polish incident, which had been reported by *Flight*, when one of its contributors, C. N. Colson, pleaded in an article for Amy to change her plans. Colson recognised that the westward crossing of the North Atlantic would be the most dangerous part of the three stages planned. He implored her to crate the Dragon and proceed by sea, and *then* attempt the long-distance record to Baghdad. However, his plea received a cool response from the Mollisons, who pointed out that such a change would cast serious doubts upon their aircraft's capabilities and also undermine the public's confidence in the future of aviation.

Maybe Amy and Jim took the journalist's comments more seriously than they cared to admit, since it is known that they both made out their wills just after Colson's warning. A trust fund was to be set up that, in the event of their deaths, would provide for a sum of money and a trophy to be presented annually to commemorate the outstanding flight of the year.

It is difficult for seasoned air-travellers today, when an Atlantic crossing is almost an everyday event, to appreciate fully what was at that time considered to be an almost suicidal undertaking. Neither Amy nor Jim rated their chances very highly. It was noticed by many people, Bill Courtenay amongst them, that Amy was particularly worried on the eve of the flight about their survival. If Jim managed to hide his forebodings under the haze of a cloud of alcohol, then perhaps we should at least understand his reasons for doing so.

The flight was planned to take off from Croydon airport on the morning of Whit Monday, 5 June, but was cancelled at the last moment due to unfavourable conditions. The Mollisons retired to the Grosvenor to await an improved weather report, which eventually came through during the early hours of the Thursday morning of the same week. They dressed hurriedly and were driven by Bill Courtenay to arrive at the airport at 4.10 a.m. An hour later the *Seafarer*'s two Gipsy Major engines were swung into life and warmed up, as well-wishers and a *Movietone* newsreel van waited for their departure.

The Mollisons with their DH Dragon, Seafarer, at Croydon in June 1933. (Author's Collection)

The couple exchanged last-minute farewells with their relatives and friends before crawling under the cabin's fuel tanks to the cockpit. The chocks were waved clear and, with Jim at the controls, the aircraft taxied out to the northernmost edge of the airfield in order to gain the maximum distance for the take-off run. He had now manoeuvred into the area beyond the boundary lights, not normally used by aircraft, and failed to take into account a drainage gutter which lay on the south side of the old hangars. As they turned to face a 1500-yard stretch of open grass, the throttles were opened and the heavily laden Dragon began to gather speed, but it had not travelled more than 200 yards before it struck the hidden obstacle. The sudden impact caused the port undercarriage to collapse, and as the aircraft slewed round, the remaining undercarriage leg also tore away, leaving the *Seafarer* flat on its belly and facing the wrong direction.

For one breathtaking moment every eye was fastened on the crumpled biplane, expecting that it might burst into flames at any moment and trap the flyers before they had time to escape. Almost immediately the clang of an ambulance bell rang out as the aerodrome's blood-wagon and fire-tender raced towards the damaged machine. First on the scene were Bill Courtenay and the airport's senior officer, Captain Jeffs, by which time the badly shaken Jim and Amy had crawled out from the rather bent aeroplane. Peter Masefield, a young man working at the airport at the time, remembered seeing Amy scrambling out of the *Seafarer* in tears, whilst press photographs taken soon after the accident show the disconsolate pair amidst a crowd of onlookers, staring in utter dejection at the twisted propellers and sagging wings. Bill Courtenay recalled how deeply the couple felt the disappointment, and how a very tearful Amy sat in the rear of his car holding hands with her husband as he drove them back to their hotel.

The Mollisons were not the kind of people who could easily be put off by such a setback. If it did anything at all, it only seemed to heighten the public's interest in them and to spur them on to try again. However, there was one important lesson to be learned from their mishap, and that was the need to find a less restricted area for their take-off than Croydon. During the three weeks it took for their damaged aircraft to be repaired, they searched for a more suitable venue, and finally decided to use the beach at Pendine in South Wales.

It was during this interval that the couple returned to Hull to be present at Mollie's wedding. She was now twenty-one, and about to be married to Trevor Jones, who was now the deputy Town Clerk to Blackpool. It is a sad but undeniable fact that a famous relative can often overshadow another member of the family, and never more so than on an important occasion such as a sister's wedding. Cine film taken at the time by Amy's father show the immense crowds who lined the streets outside the church, not so much to see the bride and groom, but to have a glimpse of the Mollisons. Poor Amy was in a 'Catch 22' situation. She had not only caused acrimony in her family by not inviting them to her own wedding, but she must still have harboured regrets that she had not attended Irene's ceremony. To have stayed away from Mollie's big day would only have exacerbated what for her parents were still fairly raw wounds.

When the *Seafarer* finally flew along the 7-mile expanse of beach to land at Pendine early in July, the Mollisons were astonished to see thousands of people thronging the golden sands, waiting to greet them. As soon as the aircraft rolled to a standstill on the smooth, hard surface, they were mobbed by autograph hunters and swarming crowds seeking to touch them. Only one local policeman was in sight and the newspapers reported women and children fainting in the stampede that followed the two aviators along the beach. Holidaymakers were camping out overnight on the sands in tents or in their cars, believing that a take-off was imminent the following morning.

They completely failed to understand why the Mollisons did not depart when the weather at Pendine seemed so perfect, little did they realise that good weather locally did not necessarily mean that conditions over the Atlantic were acceptable. Meanwhile, Jim and Amy were holed up in the Beach Hotel, only to find that the privacy and rest they needed before the flight were denied them. They found it impossible to sleep for the noise of the crowd who stood chattering underneath their bedroom window for most of the night. Their every move was closely monitored, even to knowing the exact time when they went to bed by observing when the bedroom lights were switched off.

The couple were to spend eleven frustrating days waiting for a favourable weather forecast, intermittently flying their aircraft off the beach to Cardiff's airport at Pengam Moors for storage overnight, and then back again in the morning. Finally, they gave up and took a train to London and returned to the Grosvenor, whilst their machine was flown from Cardiff back to Stag Lane. It was the evening of Friday, 21 July that they were alerted by the Air Ministry that there was an envelope of good weather over the North Atlantic for the next few days. The report showed that surface winds would be no more than 15 m.p.h., which indicated that at their intended night-time height of 2000 feet, the winds would be approximately 25 m.p.h., low enough not to sap their fuel supplies. It was just what they wanted to hear, and urgent calls were put through to the Shell representative for fuel and supplies to be on site at Pendine, ready for a noon take-off on the following day.

The *Seafarer* touched down on the beach during the early morning, this time with Bill Courtenay on board. The trio then took breakfast in the small Victorian sitting room of the Beach Hotel, whilst 415 gallons of fuel were pumped into the aircraft's tanks. Because of the undercarriage failure at Croydon, Jim decided to take 35 gallons less than they had on that occasion. Unfortunately, it proved to be a serious error of judgment on his part, one which was to prevent them reaching New York non-stop.

The press were on hand amongst several hundred spectators waiting to watch the take-off, and above the roar of the engines Jim was heard to

225

comment, 'If we get safely through this one, I shall retire to some cottage in the country.' Then, with a slight pause and a grin, he added, 'And look twice before I cross the road.' Amy stood alongside him, immaculately dressed in a white flying-suit, her hand vainly attempting to restrain her long blonde hair from streaking back in the slipstream of the propellers. She looked relaxed and confident as she climbed into the rear of the aircraft and her final comment to reporters was, 'Now for the greatest adventure of my life.'

Just before noon the *Seafarer* fanned a great cloud of sand and grit back from its prop-wash as it started its run down the beach. It had been closely combed to make sure that there were no sharp stones or flints that could slash a tyre, a worthwhile tip that Jim Mollison had gleaned from his fellow Scot, Sir Malcolm Campbell. After a distance of 900 yards the contour of the tyres resumed their unladen shape as the control column was pulled back and the machine's wings lifted 5300 pounds into the air. Four photographic and escort planes followed in its wake as it headed towards Tenby. The weather report had predicted thick, low cloud and drizzle over southern Ireland, and it was not long before the Mollisons began to lose sight of their escorts. It had been their intention to accompany the couple as far as Mizen Head, but with the cloud base down to almost sea level and Jim holding the machine just 20 feet above the waves, they quickly lost contact.

The Mollisons head toward the North Atlantic and their New York destination as they leave Pendine Beach on 22 July 1933. (Author's Collection)

The flight might well have come to a spectacular end in a ball of flames less than two and a quarter hours after leaving Pendine. Whilst they were somewhere near the southern-most tip of the Irish coast and flying purely on instruments through thick cloud, Amy suddenly shouted, 'Look out! Keep left.' Jim made an immediate steep climbing turn to port as they missed the edge of a cliff by only a few feet. The aircraft punched its way through a layer of nimbostratus and they suddenly emerged into bright sunshine, where they found themselves flying on top of a white carpet of cloud under a canopy of blue skies. For the next twenty hours they were to glimpse the sea only at sporadic intervals, and then for no more than an hour in total.

The *Seafarer* settled down to a steady 100 m.p.h. with 2000 r.p.m. as pilot and co-pilot accustomed themselves to the even beat of the two four-cylinder Gipsy engines. Both knew that the Dragon was incapable of remaining airborne if one of the engines should fail. Even with a normal, maximum fuel load of 60 gallons the DH 84 was unable to maintain height on one engine, let alone with a fuel overload of 355 gallons. Under such conditions it meant that with two engines, they were no more secure than in a single-engined aircraft. In fact the risk of failure was now doubled. They carried neither parachutes nor liferaft, for they knew that such items would prove useless in an emergency over the Atlantic. It was no flight for the fainthearted.

As they chased the sun westward through an extended day, they relieved one another at the controls at agreed intervals and crawled back underneath the fuel tanks to the rear of the aircraft. There were two tasks to be performed whilst 'off duty'. One was to note the fuel gauge readings on the three cabin tanks, and the other, equally important, was to verify the accuracy of the two cockpit compasses with another compass fitted in the rear section of the aircraft. A sealed barograph with a fifty-nine hours' duration chart was also carried, which would verify that the machine had not landed at any intermediate point during the long-distance record flight. The rearward trips were welcomed because it enabled them, with their small stature, to stand up and stretch their legs.

They always needed to keep their nerve when any one of the three main tanks ran dry, because at that instant the engines would falter momentarily. It was essential at that moment to make sure that the tank ran quite dry before switching to the next one, since the accuracy of the mechanically operated gauges then currently in use was not to be relied upon. Another difficulty was that due to engine noise it was not possible to carry out normal conversation and when communication was needed, it was written notes.

As the sky began to darken, the temperature fell and Jim handed the controls over to Amy as he went back to don a warmer flying-suit and fur-lined boots. When he crawled into the cockpit, still struggling to get his arms into the sleeves, Amy gave a chuckle. He was attempting the impossible – he was trying to get into her smaller suit. Throughout the night they took hourly shifts at the controls, interspersed with vain attempts at sleep on the camp bed by the front tank.

Jim had brought a novel with him to read by torchlight, but later admitted that in the end he had been far too worried to concentrate. All he did know was that the night was interminably long. There is every possibility that he relieved the boredom with more than a few swigs of whisky during this period of the flight. In his book *Playboy of the Air* he wrote:

> Flying the ocean in the more difficult direction with my wife as companion remains an outstanding memory I shall carry to the grave. As a flying companion I would still choose Amy as above reproach. Our senses were dulled. We drank black coffee to keep awake, swung legs and arms as best we could to keep warm. 'Was it a hundred years ago we walked safely and quietly on dry land?'[18]

The great danger throughout the flight was that they were unable to check any drift of the aircraft, which could lead them dangerously off course. The waves beneath them were hidden by unbroken layers of cloud and so denied them of their one valuable source of verification. Neither was skilled in the art of astro-navigation, where they might have been able to use a sextant as a check on their position. They carried no radio, and radar was still to be invented. The best they could do was to estimate their drift from the direction and speed of the wind given in their weather report, which was not very accurate in those days.

It was approximately 9 a.m. British Summer Time and they had been in the air for twenty-one hours when the sun began to rise on the horizon directly behind them. If the wind speeds had been as forecast, then they should see landfall within the hour. Suddenly, Amy passed a note to her husband; on it was scribbled one word, 'Icebergs'. Jim looked down through a gap in the clouds at the yellow-white mountains and shouted, 'We're too far north, we must have drifted off course!' At least the appearance of these floating giants confirmed that they were somewhere between Newfoundland and Labrador, but exactly where, they did not know.

They flew on for the next two hours with Jim feeling all the time that he should be steering a compass course in a more southerly direction. Fortunately, he resisted the temptation. To have done so might well have meant meeting a watery grave, for they would then have been flying parallel to the coastline but a hundred or more miles from it. Just before noon they sighted a flock of birds wheeling beneath them and it was not long before Jim shouted that he had seen land. Peering through the screen, Amy could just make out a dark promontory of land, its rock-bound coastline reaching out as if to greet them. With broad smiles on their faces they turned to each other and shook hands. They were across.

Within a short while they were able to confirm their position over the Straits of Belle Isle at the mouth of White Bay, Newfoundland. They were only a few miles north of their intended landfall and approximately 1200 miles from New York. Because of the unexpectedly strong headwinds they had encountered, some as much as 45 m.p.h., their average speed had been no more than a disappointing 87 m.p.h. It meant that they were now two hours behind their estimated time of arrival of 5 p.m. Eastern Standard Time at the Floyd Bennett Field in Long Island.

The first reported sighting of the *Seafarer* came over Nova Scotia when they flew over Mollison Field, the Canadian airfield that had been named after Jim in honour of his solo crossing twelve months earlier in *The Hearts Content*. Max MacLeod was a thirteen-year-old visiting the aerodrome that day and he recalled:

> It was a hot Sunday afternoon and the local flying instructor, H. O. 'Hump' Madden [at one time private pilot to the Duke of Windsor], was flying joy-riders at $2.00 a piece, when off in the distance a larger plane was seen circling over Northumberland Strait, some 20 miles away. Even then I knew who it was and the instructor rushed his Fleet Finch into the air and headed towards the unidentified plane. In the meantime, Jim Mollison had spotted the field and roared past at about 100 feet in the air *en route* for New York. Amy was sleeping at the time. I will always remember the thrill of seeing that black Dragon pass so low overhead.[19]

By this time the Mollisons knew that it was touch and go on whether they would make their destination non-stop. The headwinds had taken their toll on their fuel supply and in order to conserve what was left, they dropped their cruising speed down to no more than 75 m.p.h. as they headed out into the fog banks of the Bay of Fundy. At 4.32 p.m. EST the *Seafarer* was sighted over Bar Harbour as it flew down the eastern coastline of the United States. By 7 p.m. they were sighted over Boston by two local aircraft, one of which was being flown by John Polando* with four newspaper photographers on board. They flew alongside the Mollisons' aircraft and reported that Amy was at the controls with Jim seated beside her.

At this point Amy was imploring her husband to land at Boston for more fuel whilst there was still daylight available, but he was insistent that they make their destination non-stop. There is no doubt that there was a heated argument going on in the cockpit and that Amy's judgement ultimately proved to be the sounder of the two. She felt that it was too dangerous to

* John Polando had a particular interest in the Mollisons' flight for he had crewed with Russell Boardman in July 1931 to set a long-distance record of 5012 miles.

attempt a night landing in a built-up area such as Long Island with fuel reserves running as low as they were, and with an air of resignation she passed the controls over to her husband. It was not long before he found that every time he eased the control column back and the nose of the aircraft was lifted, however slightly, the engines would begin to fade. How he must have regretted not taking on the extra 35 gallons of fuel at Pendine.

It was approximately 9.30 p.m. (3.30 a.m. BST) when they sighted the red and green boundary lights of the Lordship airfield at Bridgeport, just 55 miles from their target. There was little they could do but attempt an emergency landing. By now they had been in the air for an incredible thirty-nine hours without sleep, and both were fatigued to the point of exhaustion. The airport manager had been alerted some hours previously that they would be passing overhead on their way to the Floyd Bennett Field, and when he heard their aircraft circling overhead he ordered the airport's rotary beacon light to be switched on.

The lights of the airport only seemed to confuse Jim Mollison, much as they had done at Cape Town a year earlier. Although one can only conjecture at this point, there is every likelihood that he had been using alcohol as a stimulant during the flight. In spite of several attempts by the airport manager to help the couple to land safely by directing a powerful beam onto the 3000-foot runway and by sending up one of his instructional aircraft to guide them down, they continued to make several low passes without attempting a landing.

Ever since Jim Mollison had witnessed a colleague burn to death trapped in the blazing wreckage of a Sopwith Snipe, he had harboured the fear of a similar fate. The memory must have come flooding back to him as they made their final landing approach, for just as Amy tightened her seatbelt, he undid his. He recalled the immense bravery of his wife as she prepared herself for what was to come. According to him, as they exchanged glances moments before the sickening crash, she gave him a brave smile. Maybe this was no exaggeration for she had told the press just before the commencement of the flight:

> It is not death that I fear, but partial failure. Both Jim and I would rather die than face that. After all, we should be together, and that makes the possibility of death easier to bear for both of us. It is my fate to make this flight with my husband. I want to do it, and that is the end of it.

According to one eye-witness, the unlit *Seafarer* barely cleared the telegraph wires alongside the road before it dropped towards the tall saw-grass in the marshes adjoining the edge of the airfield. Although they had overshot the runway, they might still have made a safe landing had their course been 30 yards further to their left. Instead, the Dragon bore on through the tall grass until its wheels struck the embankment of a drainage dyke. It somersaulted onto its back and slithered to a halt some 300 yards outside the boundary of the airport.

Jim Mollison was immediately knocked unconscious and thrown through the windscreen, where he was left dangling head down in the marshy slime. Amy was left hanging from her seatbelt upside down with her head in water, but still conscious. Without realising that they had landed in a swamp, she imagined for one awful moment that her hair was being wetted by petrol leaking from the fuel tanks. Not surprisingly, she later described the feeling as 'the most unpleasant experience of my life'. Fortunately, she suffered no more than a few cuts and bruises as she was slammed against the side of the cabin, and was able to crawl out over the tops of the fuel tanks and climb down into the marsh. She then managed to stagger round to the front of the wreckage and pull her husband clear of the machine.

They were both extremely fortunate to have survived the crash on two counts. First of all there had not been a fire or explosion (Jim had the presence of mind to cut the ignition switches as soon as the wheels touched down); secondly, the waters of the Housatonic river and the Long Island Sound were on an ebb-tide, which meant that the swamp bordering the airfield, into which they crashed, had no more than a few inches of water in it, otherwise they might both have drowned.

Rescuers were quickly on the scene to find Amy nursing her injured husband and moaning, 'He couldn't see, he couldn't see.' As they were being borne away on stretchers, hastily made from lengths of fabric torn from the aircraft's wrecked wings, souvenir hunters were already crossing over from the road onto the marsh. It is estimated that within two hours the *Seafarer* was literally ripped apart. Pieces of struts, landing wires, instruments, nuts and bolts, in fact anything that was removable, was taken. What parts of the aircraft could not be removed had graffiti scrawled on them, some with messages that were none too complimentary.

Amy and Jim were both taken to the local hospital, where Jim's face needed thirty stitches to close a gash that ran from his left eyebrow down across his cheek and almost to his mouth. In addition, he required a further seventy stitches to other wounds on his arms, body and legs. According to one newspaperman on the spot, a rather pale-faced Amy sat in her mud-stained white flying-suit, nervously smoking a cigarette, whilst her wounds

An aerial view of Seafarer *on the morning after their crash at Bridgeport, Connecticut.* (Harvey Lippincott)

were being dressed. He was slightly amused to notice the incongruity of the solitary turquoise ear-ring which dangled from her left ear lobe, and the fact that her orange-painted fingernails were tinged with blue slime from the marsh.

During a brief interview at the hospital the following morning a heavily bandaged Jim Mollison made no attempt to make excuses for the crash. The *Bridgeport Post* quoted 'Captain' Mollison as saying, 'The crash at Bridgeport I attribute to fatigue . . . I was unable to judge my distance properly.' Similarly, the *New York Times* featured their pictures with a leading article in which he admitted, 'I was so tired that I couldn't tell where I was putting her.' In spite of the ignominy of their arrival, no one could deny the achievement of their North Atlantic crossing, and the praise they received was unstinting.

Much against the advice of the hospital doctors, the Mollisons were insistent that they leave for New York the day after the crash and they were driven by ambulance back to the airport, where one of the Shell Oil Company's amphibian aircraft was waiting to fly them to the Floyd Bennett Field. Jim was carried to the plane on a stretcher, whilst Amy was lifted from her wheelchair and into the aircraft's small cabin. As the red and yellow amphibian took off, it was met in the air by three escort aircraft flying in an impressive formation. The lead plane was flown by Commander Frank Hawks, a man famed for his record-breaking transcontinental flights in the late twenties and early thirties, and who introduced Amelia Earhart to aviation.

Sikorsky S-39B amphibian, NC-58V, of the Shell Petroleum Corporation arrives at Floyd Bennett Field Airport on 24 July 1933 with the injured Mollisons as passengers. They are followed from the runway by the Sikorsky amphibian New York American; *then Frank Hawks in his Northrop monoplane, and finally by a Fairchild aerial survey aircraft.* (Shell Archive photo via Hugh Scanlan)

When they landed at the Long Island aerodrome, a police motorcycle escort of ten riders sped out to meet the Shell amphibian as it was guided onto the concrete apron in front of the main hangars. Newsreel cameramen were poised on the roofs of their vans just behind the barriers, where a thousand or so had gathered to welcome the injured couple. As soon as the aircraft's propeller came to a stop and the cabin door was opened, a loud cheer rang out from the crowd. Only when the Mollisons were lifted gently out and carried to a nearby sedan did the cheering hush into a near silence. There was a pause and then the plaintive sound of the bagpipes was heard as a group of Scottish expatriates began to play 'See the Conquering Hero Comes', followed by 'Land of Hope and Glory'. The onlookers were amused, but no doubt puzzled as to why the couple had not been kept in hospital for at least two or three more days. Little did they realise that the Mollisons were determined not to be kept waiting for their moment of glory one day more than was necessary.

Over the next few days Amy and Jim received a stream of visitors to their rooms in the Plaza Hotel, mainly from the world of aviation, and amongst them Amelia Earhart. They had both met the tall, slender aviatrix after her arrival in England a year earlier, and without any hesitation she invited them to spend the following weekend with her and her husband, George Putnam, at their imposing mansion home at Locust Avenue, Rye, in Connecticut. It was an invitation that they willingly accepted.

Apart from being memorable for the Mollisons as they celebrated their victory in being the first persons to fly non-stop from the UK to the USA, that weekend coincided with their first wedding anniversary. They were asked to broadcast over NBC, with a link-up to the British Empire via the BBC, and to give an account of their flight. At the same time they were delighted to receive an official confirmation from Lord Wakefield that he was prepared to put up the cash to replace their damaged aircraft, so that they could continue with their proposed flight to Baghdad. One further surprise came when the couple received a bouquet of flowers from President Roosevelt, together with a message of congratulation, part of which read, 'As time passes, your courageous achievement will be engraved in the annals of aviation, and in that there is satisfaction.' The Mollisons were even more delighted when the message from the White House was followed by an invitation to be the guests of the Roosevelts for a Sunday luncheon in their home at Hyde Park.

It was a rare privilege for anyone to be photographed at the presidential home on a Sunday, as it was normally FDR's inflexible rule not to allow it. However, pictures taken at the time show Amy and Jim, still bearing traces of their mishap, standing alongside the three taller figures of the Roosevelts and Amelia Earhart. After lunch they were invited to swim in the president's private pool, and were quite surprised to watch the president move so effortlessly through the water, despite being completely paralysed from the waist down.

*Jim and Amy Mollison are given a ticker-tape welcome in New York on 1 August
1933 after being the first to make a non-stop flight from the UK to the USA. Notice
the Union Jack.* (The Amy Johnson Estate)

The highlight of the Mollisons' stay in the USA came on Tuesday, 1 August,
when they were given an official welcome to New York. Jim always said that
New York City welcomed the pioneer aviators with a frenzy unmatched by

any other city in the world. When Charles Lindbergh returned from making his epic solo flight in 1927, his welcome even surpassed the celebration of Armistice Day at the end of the First World War. Between three and four million people thronged the streets on the day Lindbergh was fêted, and it cost the city $16,000 to clear up the residue of the confetti and ticker-tape that had rained down from office windows.

It was customary for those to whom homage was to be paid to be met by the chairman of the Mayor's reception committee at their hotel, and be taken to an area known as the Battery at the southernmost tip of Manhattan, from where the parade always began. The timing was critical, for it was set to coincide with the commencement of the lunch hour when the majority of office workers would be out on the streets. There would then be a slow procession through the canyon of skyscrapers to City Hall where the official welcome would be given.

Jim Mollison likened the Mardi Gras type welcome that he and Amy were given on that sweltering afternoon, to 'a Caesar bringing home to Rome a new Empress'. Motorcycle patrolmen with the sirens of their Harley-Davidsons wailing, preceded the cavalcade as they tried to assist 300 policemen in keeping spectators back on the sidewalks. These escorts were followed by rows of horse-mounted police, riding twelve abreast and leading several district police bands, each blaring out their music at the head of a column of gleaming automobiles.

The jubilant Mollisons – Amy looking very *femme* in an eye-catching black beret and an attractive, matching crepe outfit, with a plaid organdie bow tied at the neckline – sat laughing and waving from the tonneau of the first open-top vehicle, as a blizzard of ticker-tape fell from the sky. Looking up they could see the Union Jack draped in their honour, as office workers, many of them sitting perched on window-sills, unleashed a summer snowstorm of paper onto an estimated crowd of 200,000 in the streets below. They were eager for a look at the first man and wife team to fly to their country from 'across the pond', and they greeted them with enthusiastic shouts of 'Attaboy, Jim' and 'Hiya, Amy.' Heroes of the day were not that uncommon, but heroines were quite a different matter.

All thoughts of the disastrous landing at Bridgeport a few days earlier were for the moment forgotten; this was the couple's moment of triumph and they savoured it to the full.

Chapter fifteen

DISENCHANTMENT

Many people have suggested that Jim Mollison became jealous of the way that America took Amy to its heart, and that this led to tension in their marriage. She most certainly did outshine her husband in the public's eye, in spite of the fact that he had been lionised in America a year before their *Seafarer* crossing, and there does appear to have been a degree of disenchantment in the marriage soon after their arrival. Amy no doubt blamed him for the Bridgeport crash, since she believed that it could have been averted had he listened to her and put down at Boston to take on more fuel. However, whatever their personal feelings may have been at this time, they both knew it was imperative that they kept the image of a happily married couple if they were to maintain a lucrative flying partnership.

One wonders just how much Amy trusted her husband as she bade him farewell on the New York quayside on the Friday evening of 11 August, when he embarked on the White Star liner *Majestic*. She knew it was necessary for him to return home to supervise the rebuilding of a replacement Dragon, but she also knew, as did most of her friends, that he had a weakness for women. Maybe she was too preoccupied with the prospect of her invitation to be the guest of the Putnams that weekend to allow such thoughts to linger in her mind for too long. Amelia Earhart, as the older woman, became a role model for Amy, and it seems likely that the example of her independent lifestyle ultimately became a factor in the Mollisons' estrangement.

Apart from making record-breaking solo flights, her hostess was also doing all the things that Amy longed to do, from fashion design and

237

modelling to being the aviation correspondent for glossy magazines such as *Cosmopolitan*. What was more important, she had placed her foot firmly on the ladder in commercial aviation – something Amy particularly envied – by becoming the first woman vice-president of an airline. She was also an aficionado of fast cars at the time of Amy's visit, driving a sleek little 1932 Essex Terraplane coupé in which she was tempted to drive faster than she should whilst Amy was her guest.

She had driven Amy up to Boston for the day to show her around the Boston & Maine Airways Company, in which she held shares. On the way back, late in the evening, she was demonstrating the car's lively performance when they heard the wail of a police siren getting louder by the second. The motorcycle patrolman pulled in front and waved the couple to a halt. He ambled back and leaned on the driver's door to peer at the two women. The tousled-haired American sat impassively, looking straight ahead of her, trying to remain as anonymous as possible. A sly grin began to spread across the policeman's face as he recognised the woman seated behind the wheel. Slowly, he pulled out a note-pad from his top pocket, but instead of writing out a ticket, he asked for an autograph, adding, 'I've just had a bet with one of my buddies back there, that it would be you, and I'd get your autograph.' Amelia obliged without batting an eyelid. 'Thank you kindly Ma'am. Do drive carefully,' he said.

The two women proceeded on their way giggling quietly to themselves. Little did the impudent patrolman know that he might well have had two famous signatures in his note-pad that evening.

Amy's stay in the United States coincided with the arrival of a formation of Italian flying boats, carrying 115 members of the *Regia Aeronautica*, a fleet led by their air minister, General Italo Balbo. It was the tenth anniversary of Mussolini's coming to power, and his way of showing his military prowess was to send twenty-five of these aerial giants across the North Atlantic, where all but two of the Savoia-Marchetti machines arrived to attend the 1933 Chicago World's Fair. They stayed for ten days before recrossing the Atlantic via the Azores, to arrive over Rome and land as victorious heroes on the river Tiber. It was an impressive display of aerial might, which ultimately deluded the Italian dictator into believing in his own invincibility.

Although Amy made little contact with the bemedalled and ostentatious Balbo, other than an exchange of congratulatory cables, she did visit Chicago a few weeks after the departure of the Italians, and stayed in what was termed 'the Balbo suite' of the Drake Hotel. On this occasion she came not only to visit the fair itself, but also to attend the air races during the weekend.

Amy had a particular interest in these air races for she knew that women would be participating, much as they did at home, but the American women flew much more powerful machines. Two of the most potent racing aircraft at that time were the Wedell-Williams and the Granville Brothers' machines, the latter being known as Gee Bees. In both cases the philosophy of their designers was to pack as big an engine – including radial engines as large as 600 horsepower – into as small a space of fuselage and wings as possible, with the result that these fat, barrel-shaped racers were highly unstable. The race circuit was marked out by a series of pylons, which meant that those competing were known to the crowds packing the grandstands as 'pylon polishers'. Speeds of 250 m.p.h. which might not seem high by today's standard, but for their time they were exceptionally fast, meant that these pilots could be pulling six or seven 'g' on a pylon turn. With four pylons on a circuit and a race of as many as twenty laps, they could encounter up to eighty turns in one event.

Whilst Amy was present for the Frank Phillips Trophy event, one of the well-known women competitors, Florence Klingensmith, was killed. As she rounded the home pylon at less than 50 feet and climbed away on her next lap, some fabric began to detach itself from the port wing of her Gee Bee racer and the debris flew back onto the tail surfaces, jamming them. The unfortunate pilot attempted a parachute jump just before the machine nose-dived into the ground, but she was found with the tangled harness of the 'chute wrapped around her.

This tragedy may well have influenced Amy's atttitude towards the use of the parachute as the means of escape from a crippled aircraft, for when she was asked a few years later, 'What are you most frightened of, Amy?' her reply was, 'The one and only thing I am terrified of is a parachute descent. Even if the wing falls off an aeroplane, it still feels like solid earth to me. Although I have to wear a parachute, I know inside me that I could never, *never* pull the cord and trust myself to the air.' One needs to keep this statement in mind, since it might well have a bearing on solving the mystery of Amy's own death.

Another contestant on that summer's day, with whom Amy eventually became friendly, was the diminutive flyer, Mae Haizlip. Like Amy herself, she was part of a man and wife team, which had carved its own niche in transcontinental and high-speed pylon racing in the USA. Wearing her white gaberdine flying helmet, the sultry-eyed beauty epitomised the glamour of the 1930s. Her equally short-statured husband, Jim Haizlip (he was under 5

The Mollisons with the US record-setting, racing pilots Jim and Mae Haizlip in 1933. (Author's Collection)

feet tall) had taught his future wife to fly whilst running his own flying school in Oklahoma, and between them they managed to capture many of the speed records of that period. Mae was once quoted as a saying, 'I may have been the only bride ever to learn about racing airplanes in the bedroom. Using a length of string looped around our bedpost, Jim would go over his strategy clearly and precisely, using the bedpost as the pylon.'[20]

Amy must have envied the fact that women seemed to play a much more important part in aviation in the States than in her own country. By 1930 there were approximately 200 licensed women pilots, who were not only employed as sales representatives for some of the smaller aircraft companies, but were beginning to gain posts as transport pilots, flying instructors, aerial crop surveyors, and were even gaining executive positions within the industry. One of the driving forces behind their advancement was the formation of a group of women known as the 'Ninety-Niners', the first women pilots' organisation of international importance, so called because at first it was thought that there were only ninety-nine licensed female pilots eligible to join it. Amongst its founder members were Amelia Earhart, who was eventually elected President, Pancho Barnes, Louise Thaden and Bobbi Trout.

The catalyst for the formation of the 'Ninety-Niners' had been the 1929 Women's Aerial Derby, an event in which twenty women entered for an eight-day race across America, from Santa Monica to Cleveland, Ohio. Fourteen completed the race, which came to be known rather derisively as the 'Powder Puff Derby'. It was an event in which women were allowed for the first time to compete as part of the National Air Races. There was much male opposition at the time to women participating in transcontinental racing, an attitude which was summed up rather neatly by one of the more blinkered reporters, who said, 'Them women don't look good in pants.'

Amy flew back to Manhattan after the weekend to accept an offer arranged by Amelia to spend a brief spell as a co-pilot with Transcontinental and Western Airlines (TWA).* The airline was anxious to attract more women passengers and they jumped at the opportunity to gain publicity by using Amy to do so. The message they wished to put across was that flying is safe. They felt that if they could get women to travel by air, then the men would soon follow. Amelia turned up to be present for the photo-opportunity at the Newark airport on the morning of Amy's departure and to pose alongside her English counterpart. It is interesting to note that as the couple stood together by the shiny corrugated body of the 'Tin Goose' (as Ford Trimotors were familiarly known) to be photographed, Amelia kept to the right of Amy, the ploy she had learned from her husband, to ensure that she would always get first billing in any caption to the picture.

Amelia Earhart and Amy Mollison pose for the camera beside a TWA Ford TriMotor, 5-AT, passenger airliner at Newark Airport in September 1933. Amy carried out the normal duties of a co-pilot on this aircraft for a brief spell. (The Amy Johnson Estate)

* Amelia had been the assistant to the General Traffic Manager of Transcontinental Air Transport, which was the foundation company of TWA.

TWA was one of America's main airlines and it operated a regular service between New York and Los Angeles, on it Amy, now dressed in a neat blue uniform and wearing a cute little beret with the TWA badge, was to gain valuable flying experience. According to her book, *Skyroads of the World*, there were several things about American commercial aviation in 1933 that favourably impressed her. The first was that there was less discrimination against women, and that a person was judged solely on ability and qualifications rather than on their gender. She also felt that commercial flying in the States was more fully developed than in any of the other countries she had visited, one of the factors being that it was such a large country, with vast distances between cities. Furthermore, it then had a population of 126 million, which made it worthwhile to run such services, and the Bureau of Air Commerce had, since 1926, not only subsidised the airlines, but also provided the finance for the advanced ground equipment necessary to run them.

During the forty-eight hours that Amy was with TWA, she was required to carry out the full range of duties of one of their airline co-pilots, and, she was determined to learn all she could. With the old, three-engined Ford airliners, cruising at 110 m.p.h. on a 2500 mile journey, it took a full twenty-four hours to cross the continent. There were fourteen scheduled stops along the way, with a change of aircraft at Kansas City. Passengers could then either stay there overnight or proceed on a night flight. Crews were changed at every quarter section of the journey. For Amy, this meant not only long spells in the cockpit, but also being required to supervise the loading and unloading of mail and baggage along the route, as well as arranging for the seating and comfort of the fifteen passengers during the flight.

It did not take her long to realise that America was years ahead of other countries with respect to aviation, inasmuch as their airlines were using quite sophisticated navigational aids, which enabled safe night flying. Unlike her other long-distance flights, where a pilot relied entirely upon a set compass course, she could on this flight wear headphones in the cockpit which sounded a steady purr, telling her that she was on course. If she wandered to one side or the other off the correct line, a change in note in the headphones immediately warned her. In addition, a weather report was received every hour from a second radio set, and by using a switch-over key she could speak to ground stations or other aircraft. In cloud or fog the aircraft's sensitive altimeter, graduated for each 100 feet by a sweep of its second hand around the dial, and its artificial horizon and gyro-compass, made a pilot independent of visual contact with the ground. Night flying was further assisted by flashing beacons placed along the route, with special ones beaming a morse code to indicate aerodromes. It all added up to the luxury of a land of modernity.

Amy broke her journey at Kansas, where she was invited to tour the maintenance sections of the airline's repair shops before flying on to Los Angeles. Whilst she was in California she took the opportunity to visit the Douglas Aircraft Company's plant in Santa Monica, and was able to inspect their new DC1 (Douglas Commercial No. 1) twin-engined airliner, the forerunner of the ubiquitous wartime Dakota, aptly named 'the Volkswagen of the skies'. It must surely have disheartened her when she compared this smooth-skinned, all-metal monoplane, with its two powerful 700 h.p. Pratt & Whitney radial engines, to the obsolescent de Havilland design in which she was now expected to risk her neck by flying back across the North Atlantic. A letter to the editor of the *New York Times* whilst Amy was in the States hinted at this very fact, when a reader cast doubt upon the suitability of the Dragon for the purposes that the Mollisons intended. Modern aircraft such as the DC1 and the Boeing 247, now coming into production, were capable of speeds in excess of 200 m.p.h. an important safety factor when crossing vast expanses of ocean. The *Seafarer* had lumbered across at an average speed of less than half of this, thereby endangering its crew unnecessarily.

When Jim returned to meet Amy in mid-September, it was to Montreal and not New York that he came, for he had been persuaded by his wife to use the 5-mile stretch of hard sand on the edge of Lake Simcoe at Wasaga Beach, Ontario, for take-off, instead of the runway at Floyd Bennett Field. Amy had decided that with the 600 gallons of fuel that they would require to make the Iraqi capital non-stop, they needed the maximum amount of space available. She also reasoned that because so many record breakers were using FBF it was losing its news value, whilst departures from Canada were comparatively rare.

It is significant to note that when Amy and Jim were reunited outside Toronto's Union station on 15 September, the press were not slow to point out that their greeting was an exceptionally cool one. There was no embrace, just a polite 'Hello' and a hand-clasp, before they made their way to the Royal York Hotel opposite.

They wasted very little time before surveying the Wasaga site, arriving there the same afternoon, and were met by Bob Loader, the genial manager of de Havilland Canada, and his colleague John Adamson. It had been Adamson who had suggested the Wasaga beach to Amy, after she had made enquiries about using the St Hubert airfield at Montreal during a previous visit to the city. The Mollisons were quite impressed at first by the firmness of the sands, but less so when they examined its slightly bumpy surface. As a guide to the severity of the undulations, they arranged for Bob Loader to drive them in their MacLoughlin-Buick along the 3-mile stretch at 70 m.p.h. whilst they sat in the back.

After some consultation with Loader on the time-schedule for uncrating and assembling what was to be *Seafarer II*, due in Montreal the next day, they decided that they would make their attempt on the long-distance record with the help of a full moon in ten days' time. In the meantime, they would relax on a two-day fishing holiday as guests of a Canadian friend at his Muskoka home.

Seafarer II, *G-ACJM, at Wasaga, October 1933.* (Author's Collection)

Bob Loader was as good as his word, and under his supervision the de Havilland workshops at Weston pulled out all the stops, so that within a few days the plane was ready for the Mollisons' return as promised. As the pristine, black-painted biplane was pushed out from the hangar, resplendent in its contrasting white lettering, people were quick to notice that the G-ACJM registration contained both flyers' initials. Perhaps it was a good omen – Amy and Jim certainly hoped so.

The day appointed for the Mollisons' departure came and went as the weather conditions over the Atlantic deteriorated and the take-off was postponed. These delays were grating on the flyers' nerves, particularly so for Amy, who tried to take her mind off things by spending her time horse-riding. There were several worries at the back of both of their minds. Would the aircraft be capable of lifting 7334 pounds – more than three times the weight of a standard Dragon – as the design team at Stag Lane assured them it would? They also knew that time was fast running out for transatlantic flying, now that the winter weather was closing in on them. One other worry was that when Jim made one or two exploratory take-offs from the beach in a borrowed Puss Moth, they realised that Adamson had failed to mention the strong cross-winds which blew predominantly across the beach for most of the year.

By the morning of 3 October they decided to wait no longer. They had been delayed for eighteen days since Jim's arrival in Canada, and the Dragon was now tanked up to its maximum capacity of 608 gallons and ready for take-off. Since their Bridgeport crash, a new long-distance record of 5657 miles had been set up by two Frenchmen, Paul Codos and Maurice Rossi, which meant that the Mollisons definitely had to reach Baghdad if they were to set a new record. A few days earlier, Amy had been thrown from her horse whilst taking a cross-country ride, and although her injuries had been no more than a few bruises, it probably did little for her morale at the commencement of a long and dangerous flight. She made what was to be a chillingly prophetic statement, when she told reporters, 'If we're very, very lucky, we may make it, but it will have to be before the first week in October, for we are not looking for suicide. Yes, I'll fly till I die – and I hope I die flying.'

On the first two attempts the Dragon failed to unstick, even after runs of as much as 2 miles. On the third attempt the tail rose sluggishly, and after a mile they hit a slight ridge which shot the plane into the air. Jim later told Bill Courtenay that the drift towards the shoreline was so strong that he was forced to put the overladen aircraft down with a spine-rattling thud. Inevitably, there followed an ominous crunch from the protesting undercarriage.

The air must have been blue with invective at that instant as Jim vented his feelings with some choice expletives, for he recorded that Amy told him, 'Just keep all that to yourself.' She had no doubt heard it all before at Croydon when they had suffered a similar fate. Such unnerving experiences, and the ensuing recriminations, could hardly have enhanced their marital harmony. The cruel irony of it all was that had they waited but another thirty minutes, the wind was to veer so that they would have been able to take off directly into it.

It seems that their aircraft could not have been too badly damaged, for after draining most of the fuel it was flown back to the Downsview factory. Close examination revealed that the landing gear had been slightly twisted on impact, but that was all. Repairs were quickly carried out and the aircraft was declared ready to be flown back to Wasaga for a further attempt. However, the weather broke, and four days later the Mollisons issued a statement to the Canadian press which made it clear that the flight was off. The ill-fated *Seafarer II* would be shipped back to England.

Not long after their failed attempt DH Canada successfully sued the Mollisons for garaging and assembly bills which had not been paid. Whether the unpaid bills were due to a shortage of money or to a dispute between the couple on exactly how much each should contribute to the costs, is a matter for speculation. Jim may well have blamed Amy for listening to Adamson and choosing the Wasaga site; equally she might well have countered that it was his fault for damaging the original Dragon at Bridgeport in the first place. It is doubtful whether the unpaid bill ever worried Jim Mollison, for it was well known that he had a reputation for leaving a trail of debts wherever he went. The Scottish pioneer aviator, John Grierson, once made a passing comment to air historian Richard K. Smith, in this context. The latter recalls:

> There was always an aspect to Mollison's career that puzzled me. By 1933 this man had done Australia–UK; UK–Cape Town; the North Atlantic and South Atlantic, an almost incredible score – so why didn't he get a knighthood? John Grierson told me that when JM was in the RAF on a short service commission whilst stationed in India and his term ran out, he left his station to return without paying his mess bill. As if that was not bad enough (it simply wasn't done), he later *bragged* about it, and did so repeatedly in public. Men who do this sort of thing demonstrate that they are not gentlemen and *they* don't get knighthoods.

Jim sailed from Montreal with the dismantled *Seafarer II* to Southampton, whereupon it was flown back to Hatfield to await a buyer. He immediately sailed back to meet Amy in New York, having promised her that they would spend several weeks together on holiday in Bermuda. This did not escape the cynicism of the editor of *Aeroplane*, the curmudgeonly C. G. Grey, who took great delight in 'congratulating' Mollison on being the first man to accompany a potentially record-breaking aeroplane on a double sea-journey across the Atlantic!

In the meantime, Amy had been admitted to a New York hospital for a medical check-up, ten days after the aborted take-off from Wasaga. She was described as being in a highly nervous state and after examination diagnosed to be suffering from a stomach ulcer. Knowing that her stay in hospital might be a prolonged one, it was agreed that Jim would travel on from New York to Bermuda by steamship, and she would meet him there at the conclusion of her treatment. There appears to have been some ulterior motive in choosing Bermuda for a vacation, inasmuch as it had been reported that after the failure at Wasaga, they had considered using a South Atlantic route via the Azores for their long-distance flight.

According to Jim Mollison, the time that he and Amy 'honeymooned' together in Bermuda was the happiest of their entire marriage. Amy joined him after a month in the New York hospital, and he describes how they whiled their days away on sun-kissed beaches, whilst the evenings were spent dancing to the latest Latin numbers. They were both good dancers, and no doubt they cut quite a dash as they swayed to the haunting melodies of the tangos which were then in fashion in the local nightclubs. Maybe if they had spent more time getting to know each other better the marriage might have lasted, but their social interests were not really compatible. He was a born sybarite, a man who lived for the moment without thought for tomorrow. Money came and money went. She was more cautious, for she recognised that the days of the record-chasers and easy money were fast coming to a close.

Their time together in Bermuda must have been stormier than he conveyed in his memoirs, since at the end of six weeks on the island he returned to London alone, leaving Amy to spend the next three months in Florida. Doctors had recognised that she was living on an emotional knife-edge, and in a letter to her parents she admitted that on occasions she felt her self-control slipping away and was subject to 'violent nerve storms'. This aspect was highlighted when she was pulled in for speeding during the early hours of the morning by two plain-clothes policemen, who had been tailing her in an unmarked patrol car whilst she was returning to Palm Beach from Miami. She had been partying with friends and alleged that as the police car closed up behind her, she thought that it might be a hold-up. In order to shake them off she put her foot down even harder, until eventually her Mercedes was forced to stop. Whilst being questioned she lost her temper and slapped one of the officers across the face, with the result that she was arrested on the spot. When she appeared in court a few days later, she was fortunate to suffer no more than what was for those days a hefty fine of $50, plus a further $20 for costs.

Not surprisingly, certain sections of the popular press in Britain were hinting that all was not well with the Mollisons' marriage. Jim vehemently denied any estrangement, whilst Amy was in continual correspondence with her father seeking to gauge the measure of public reaction back home about it. The couple both had a great concern for their public image as the romantic 'air lovers' at a time when divorce was not as readily accepted as it is today. They knew that it was in their interest to maintain the veneer of a happy marriage, even if it did not exist, although to be fair to Amy she admitted privately that she was still very much in love with Jim. If their marriage had been a relatively smooth and sudden coming together, then its dissolution was about to be a protracted and turbulent one.

Jim Mollison's blue eyes betrayed more than a flicker of excitement as he read an aviation trade journal in the early January of 1934. Here was the news for which he had been waiting ever since he heard of the £15,000 prize money that was being offered by its sponsor, Sir William Macpherson Robertson to the winners of a proposed England to Australia air race. The de Havilland Company was determined to be a serious contender for the prizes and were now inviting orders for a limited number of their new Comet Racer, which was now in the development stage. Jim wired Amy in Florida immediately, asking if she was prepared to go fifty-fifty on one of the three Comets that were to be built, and stressing that he needed to know quickly, as it was a question of first come first served.

With the memories of Croydon and Wasaga still lingering painfully in her mind, Amy was at first sceptical, for apart from the £5,000 needed to purchase the Comet, she was convinced, after visiting the Douglas Aircraft Corporation in California, that the Americans who would be entering had aircraft that were technically superior to their British counterparts. She knew that advanced design features, such as variable pitch propellers, retractable undercarriages and wing flaps, were currently in production in America. However, in spite of this she wired back her acceptance, for much as she was enjoying life in the States, she was homesick for England and was anxious to get back into the flying scene.

One of the first things the Mollisons did when Amy returned in mid-February was to meet with Bill Courtenay to see where they stood in regard to their contract with the *Daily Mail*. Whilst they had only completed one third of their proposed flight to Baghdad, and had failed therefore to fulfil their end of the bargain with the newspaper, they had made an historic flight, and Amy had become the first woman to fly to the USA. The newspaper was under no obligation to pay out, but according to Courtenay, who argued that they should receive at least a third of the sum due, they handed over what he considered to be 'a substantial cheque'.

Amy returned to live at the Grosvenor House Hotel with Jim during the early part of 1934, but she soon came to hate every minute of it. The enforced life of idleness, broken only by the constant round of invitations to parties put on by the *glitterati* of what was termed Mayfair's café society, was too boring and stifling for Amy. She was determined to lead a life of her own choosing, and not that of her husband. His philosophy at this time is graphically expressed in his book, *Playboy of the Air*.

> For a morning visitor to tear apart the dark curtains of my bedroom windows at 11 a.m. is for me torture. England's grey daylight spells work and offices, trams and trains, realities, punctuality, all of these things I dislike most.
>
> Life and enjoyment begin when daylight fades and the bright lights are a-twinkle. Cocktail bars and the clubs from here to Honolulu are opening their doors to the faithful. Music, wine, moonlight through the palms, beautiful gowns, beautiful women inside them, whisperings, shaded lights. These are things only of the night.
>
> Daylight comes to me as a break, an interval for sleeping until an afternoon drink that helps bring on another evening.[21]

It is not difficult to gather from the last part of this hedonist's 'ode to joy', that Jim Mollison was in the early stages of alcoholic addiction well before the age of thirty-five, and that in a much more enlightened age, friends would have warned him of the long-term dangers to his health. Whether he would have listened to their advice is quite a another matter.

Chapter Sixteen

BLACK MAGIC

G eoffrey de Havilland's decision to go ahead with the design of an aircraft capable of winning what had now become known as the MacRobertson Air Race, was an exceptionally brave one. A financial outlay of £50,000 was needed to build the three aircraft – a daunting sum of money for a small company to spend at a time when the British aircraft industry was in the doldrums. He knew that there was no way that this investment could be recouped solely by the sale of the three Comets, for it was not practicable to sell them for more than £5000 each. De Havilland was an intensely patriotic man who did not want to see the race given away to the Americans by default, and so, in the absence of of any serious interest from any of the other aircraft manufacturers, apart from the Airspeed Company, he decided to proceed from the drawing-board stage. There was prestige and a sense of achievement to be gained if the company produced a winning aeroplane, but not much else other than a boost in sales of their current range of aircraft. It was a bold gamble.

It says little for the British government's foresight and understanding of the needs of its own aircraft industry that the UK did not already possess a high-speed airliner capable of winning the MacRobertson in 1934. Unlike the United States, which could enter heavily subsidised modern airliners such as those being produced by the Douglas and Boeing companies, Britain had neglected the development of fast modern airliners on its Empire routes. Even KLM's orders for the Douglas DC2 for use on a swift passenger and airmail service between Amsterdam and the Dutch East Indies, and Holland's development of its own rival aircraft, did not alert the blinkered officialdom of the Air Ministry. Geoffrey de Havilland's warnings and the pleas of others fell on deaf ears, and the Hatfield company was left to rescue its country's prestige without the benefit of subsidy.

De Havilland had a tight time-schedule on their hands, since their design team knew that they had less than nine months in which to work from a blank sheet of paper to a successful test flight. It soon became apparent that the main design parameters were determined by the aircraft's need to cruise at 220 m.p.h., and for it to have a range of at least 2800 miles. The result was a sleek, twin-engined monoplane which was to become the seminal design for the much-acclaimed Mosquito, one of the most successful aircraft of the Second World War. It incorporated a centre-section wing flap which assisted take-off and landing approaches, and underslung engine nacelles which provided space for a retractable undercarriage, so necessary in an aerodynamically clean design.

By the end of February 1934 the directors at de Havilland had allocated all three of the Comet Racers they intended to build to those applicants most likely to bring them a victory. Apart from the Mollisons, the aircraft were to go to the sponsors, Bernard Rubin, a wealthy Australian, and A. O. Edwards, the Managing Director of the Grosvenor House Hotel. Rubin intended to take part himself and to crew with another pilot, but owing to illness he pulled out at the last minute, leaving the former Fleet Air Arm pilot, Owen Cathcart-Jones and the flying instructor, Ken Waller, to fly for him. Surprisingly, the Grosvenor House director decided to sponsor the Mollisons' old rival Charles Scott as his pilot, and for Tom Campbell Black to act as co-pilot in a Comet that was to be named after the hotel.

One can only speculate on the reason for Edwards's choice of Scott and Black rather than the Mollisons, especially as the couple were resident in his hotel. His sponsorship could have saved them finding £5,000 out of their own pockets, at a time when the pay-out from their *Seafarer* flight had not been what they expected. Although the couple were by no means on their uppers, they were not exactly flush with money either. Amy's well-stocked wardrobe from renowned Parisian couturiers and their penchant for fast, expensive cars – Amy with her Hispano Suiza and Jim with his closely guarded Buick – saw to that. Bill Renwick, an international rugby player and Jim's cousin, was an occasional visitor to the Mollisons' apartment at the Grosvenor, and he remembered that in those days 'their suite was sheer luxury, an open-house and "spooned" upon by nearly everyone'.

Whilst A. O. Edwards recognised that the Mollisons' reputation as solo flyers was unassailable, their teamwork was another matter, since their marital disputes were hardly a closely kept secret in his hotel. One well-known London actress and friend of the Mollisons recalled receiving a telephone call from a member of the Grosvenor's management, telling her that there had been a violent quarrel between the couple the previous night and that their bathroom was covered in blood from a fight. Certainly they both had quick tempers and were capable of becoming involved in a domestic slugging match. Jim could be aggressive when he had drunk too much, and Amy was no mean sparring partner when she was roused, as the Florida police could well testify.

As a consequence of Edwards's decision, plus the fact that Jim Mollison had openly announced that he and Amy did not intend making any night landings during the air race, the pundits were now backing Scott and Black as the firm favourites to win the MacRobertson. Their prediction did little to improve relations between the rival teams. Amy disliked Charles Scott intensely, since she had never forgiven him for his boorish behaviour towards her during her 1930 flight to Australia. Jim's feelings towards him were no more friendly, although less personal since they hinged much more around professional jealousy.

In spite of the preparations for the air race, Amy was not deflected from her aim of making a serious career for herself in aviation independent of her husband. Overtures to Colonel Shelmerdine, Director of Civil Aviation met with little response, so she changed tack and with the help of Bill Courtenay managed to land a job with the *Daily Mail* as their aviation editor. Her contract was only for six months, but that suited her for it expired just before the start of the MacRobertson in October. Although she was not a particularly gifted writer, she believed passionately in the future of civil aviation and was well informed on aeronautical matters, and she soon settled down to make a success of her new occupation. Not to be outdone, Charles Scott was appointed in a similar position with the Mail's rival, the *News Chronicle*, soon afterwards.

The Mollisons moved out of the Grosvenor for the summer and rented a country cottage in Lurgashall, a small village to the north of the South Downs, where their aim was to get themselves physically and mentally fit for the race. Although Amy was still in love with her restless and intemperate husband, their marriage only seemed to survive because they led separate lives. As she admitted to her father on one occasion, her dilemma was that she was as unhappy with Jim as she was away from him.

During that summer Amy sailed for America on the *Majestic* to follow up an offer made by the Beech Aircraft Corporation to become their agent in the UK. She accepted the offer and, to show her good faith and commitment, she placed an order for one of their modern staggerwing cabin-biplanes for her own personal use. Now an admirer of all things American, she combined her business trip with a brief holiday in Arizona before returning to Lurgashall in mid-August.

Whilst Amy was away Jim was free to do virtually as he pleased, and he was soon back to the lifestyle he liked best. Although Amy suspected that he was being unfaithful, she had no hard evidence of it. Jim Mollison had a certain brooding sexuality, which many women found hard to resist, and his current clandestine romance was with the blonde-haired revue actress, Dorothy Ward. Although she was fifteen years his senior, she had retained her good looks and youthful figure, which she used to good effect in her appearances as principal boy in pantomime. She was a versatile actress, who had been widely acclaimed on Broadway and on the West End stage during the 1920s. Whether

Dorothy Ward. (Author's Collection)

Amy was aware of their liaison is open to question, but by the time her husband's memoirs were published three years later, she was left in no doubt, for the book was unashamedly dedicated to the actress. According to Dorothy's son, Peter Glenville, on the rare occasions his mother did mention Jim Mollison, she always referred to him as 'that splendid cad'.

When Amy returned from the USA she was delighted to learn that Bill Courtenay's approaches to Edward Hillman to employ her as one of his pilots had paid off. Hillman, who had recently moved his airline from Maylands, near Romford, to Stapleford Tawney, had agreed to Amy piloting one of his six Dragons on the Le Bourget, Paris, run.

One young pilot with the airline at that time was the late Wing Commander G. D. 'Flip' Fleming, and he recalled Amy arriving, dressed in a smart grey two-piece suit, along with her husband. They were accompanied by a newsreel team trailing wires all over the place as they tried to organise everyone into position for their cameras. The cameraman thought that Amy

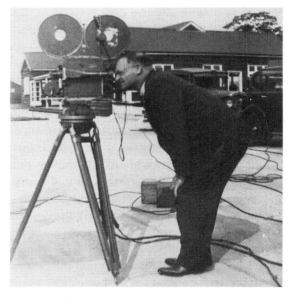

ABOVE LEFT:
Amy wearing 'Flip' Fleming's leather jacket during her visit to Hillman Airways at Stapleford Tawney on 24 August 1934. (G.D. Fleming)

ABOVE RIGHT:
Captain 'Flip' Fleming alongside one of Hillman's DH Dragon Rapides in June 1934. (G. D. Fleming)

RIGHT:
Edward Hillman takes a peep through the cine-camera which was set up to record Amy's visit to his airline. (G.D. Fleming)

looked out of character as an airline pilot dressed as she was, and asked that she be given something more suitable to wear over her outfit. 'Flip' Fleming remembers loaning her his old flying-jacket, one that he only used whilst driving his open-top sports Alvis. Unfortunately, just as the cameras began to whirr and as Amy climbed into the Dragon, she banged her head and swore, so that the whole sequence had to be retaken. A further problem arose later in the day, when several passengers refused to fly with a woman pilot. It was all a far cry from the attitude Amy had met with in the USA.

Fleming remembered that he was asked to escort Amy on her flights to Paris for a few days, and that he got to know her quite well. They would have lunch together at Le Bourget airport and sit and chat before flying back. He said:

> She had large blue eyes that flashed about in a rather nervous manner. She seemed shy and nervous in public places, but frank and sincere when we were alone. I don't think she became a famous pilot due to any natural gifts, other than perseverance and sheer guts. Technically speaking, her flying was not above average, and her landings were erratic; her navigation and airmanship were mastered by sheer hard work and not instinct, but her courage, determination and endurance in the face of these odds was all the more remarkable.

Of Jim, he had less to say, other than he thought that 'he was very sure of himself, didn't give a damn for anyone, and drank too much'.

Amy flew with Hillman Airways until mid-September but was forced to relinquish her post in order to concentrate on the forthcoming air race to Australia. She fully intended to resume employment with the airline after the MacRobertson race, but her prospects dwindled when Hillman died suddenly from a heart attack at the age of forty-five on 31 December. It was hardly a surprise, since this bustling, overweight man suffered from high blood pressure and worked over a hundred hours a week. At the time of his death his company was the largest unsubsidised airline in Britain, and Courtenay always maintained that had Hillman lived, Amy's place on the board of directors would have been assured. Within nine months of his death however, Hillman Airways had merged with two other smaller independent airlines to become British Airways.

Meanwhile, the prototype Comet Racer, the first of the three and the one which was allocated to the Mollisons, made its first test flight at Hatfield. In its day the de Havilland 88 Comet, with its maximum speed of 228 m.p.h. at sea level, probably caused as much public excitement as Concorde was to do thirty-five years later. Whilst a technical comparison may not stand close scrutiny, aesthetically the two aeroplanes had a very similar visual impact. Both in the air and on the ground the sleek lines of the Comet caused the pulse of the beholder to beat just that little bit faster. Even today, a visit to the Shuttleworth Collection at Biggleswade, where a carefully restored *Grosvenor House* is exhibited in its eye-catching scarlet livery, shows that it still outshines the rest of the aircraft on display.

As the date of the race drew nearer the Mollisons realised that they had to get down to some serious planning, for they were not particularly well prepared. During September they started a fitness programme and were to be seen cycling hard along the Sussex country lanes near Lurgashall, training for the big event. No doubt part of Amy's plan in getting her husband away from London was to wean him off the booze and into a dry period. How abstemious he became during this time is questionable.

Two weeks before the race, Amy wrote to her father telling him they had still not acquired all the equipment necessary for the flight. Apart from flying clothing, parachutes, lifejackets, emergency flares, permits and strip-maps, they still needed visas for permission to overfly nine different countries that lay along the route. One of these countries was Turkey, a country directly in the flight path of the first stage of the journey. Whilst none of the other competitors had difficulty in obtaining permission from the Turkish authorities, a ban had been imposed on Jim Mollison ever since he made an illegal landing there in 1931. The Air Ministry had written to Harold Perrin, Secretary of the Royal Aero Club, informing him that HM Government had made representations to the Turkish government in an endeavour to have the ban lifted, but without success. The letter added that serious consequences were likely if the Mollisons should make a forced landing in that country; moreover, it would be impossible for the British Government to intervene on their behalf if such a situation arose.

The Mollisons now had a hard decision to make: either avoid flying over Turkey altogether and make a detour of several hundred miles, thereby giving an advantage to their competitors, or trust that their two Gipsy engines would keep going as they traversed 400 miles of Turkish airspace. Certainly the prospect of spending the winter in a Turkish jail without legal representation was not a particularly inviting one.

The starting point for the race was to be the RAF's recently opened airfield at Mildenhall in Suffolk, where all twenty-one competitors were due for the commencement of an official scrutiny on the afternoon of Sunday, 14

October. The three de Havilland Comets were the first to appear on the horizon, to be met by strong gusting winds as they came in to land. Apart from Amy, all members of the Comets' crews had received 'hands-on' training in the aircraft, including at least one practice landing at Hatfield on the previous day.

One of the Comet's unfortunate handling characteristics was inherited from the design of its wing shape, which tapered sharply from root to tip. This caused a phenomenon known as 'tip stall' which, if not corrected quickly, gave the aircraft a tendency to drop and scrape a wing on its landing approach. It was made worse if a three-point landing was attempted, and pilots were advised to land on the main wheels before allowing the tail of the aircraft to drop onto its tail-skid.

The first to arrive at Mildenhall were Owen Cathcart-Jones and Ken Waller in their all-green Comet, and it was noticeable that they only narrowly escaped falling victim to the aircraft's peculiar landing characteristic. The next was the scarlet-red *Grosvenor House* flown by Charles Scott and Tom Campbell Black, quickly followed by the Mollisons in their all-black Comet, *Black Magic*, with Jim at the controls.

Attentive peak-capped RAF marshals ran towards the Mollisons' machine and positioned themselves at the wing tips as they taxied towards their station on the concrete apron in front of the hangar. This practice of guiding the aircraft whilst taxi-ing was especially necessary with the Comet, because of its extremely long nose housing the fuel tanks, which meant that once the tail of the machine was down on the ground the pilot was virtually blind in the straight-ahead position.

As soon as Amy and Jim clambered down from the cramped confines of the cabin, they were besieged by press reporters and autograph hunters. Jim, looking windswept and serious, was obviously showing some concern for the handling qualities of their new machine, whilst Amy, wearing a black beret which tended to enhance her blonde hair, seemed quite unperturbed by all the fuss. She was always a magnet for the press, even more so now that she was married to this wild and flamboyant husband.

The whole aerodrome was now buzzing with excitement as the remaining aircraft and their crews flew in at sporadic intervals throughout the afternoon. Jim and Amy began to survey the opposition. The glistening, silvered Dutch and American airliners aroused particular interest, for they were thought to be strong contenders in both the speed and handicap sections of the race – the event was really two races in one, with a prize for the fastest machine and another which would take into account the amount of payload carried, enabling an ordinary commercial aircraft to compete. In addition, there would be a gold medal for every pilot who reached Melbourne within sixteen days.

The two Dutch pilots, Parmentier and his KLM co-pilot, Moll, who were to fly a Douglas DC2 passenger airliner, maintained an air of insouciance, as if they were already mentally converting the prize money into guilders, whilst the ebullient American showman, 'Colonel' Roscoe Turner, laughed and joked, striking a Hollywood pose with his waxed-moustache and Ruritanian military uniform. He was there not only to win but to make sure that his Warner Brothers sponsors gained the maximum publicity. The other Dutch entry was the three-engined Pander *Postjaeger*, an airliner which was designed to give a fast service on the Schiphol–Jakarta run. Slowly it took its place in the line-up alongside the other aircraft. Its crew, the phlegmatic Jan Geysendorfer and co-pilot Dick Asjes, busied themselves with the formalities of checking in. Like their colleagues in the Douglas, they had the advantage over their fellow competitors of being familiar with the intended route, by virtue of flying regularly from Amsterdam on the Dutch East Indies service.

Finally, just before dusk, the rasping note of the Pratt & Whitney radial engine from Jackie Cochran and Wesley Smith's Granville Gee Bee erupted over the airfield. It was the last aircraft to arrive, and as it circled for a landing with its navigation lights on, all eyes were focused on it. This stubby, brutish racer had been dubbed 'the flying brick', for its landings were anything but smooth, and this one was no exception. Jackie Cochran was at the controls, and she brought the plane in with a thud that must have rattled her teeth. Some forty years later, she recalled her impressions of this particular aeroplane.

> Gee Bee stands for Granville Brothers, a Springfield, Massachusetts, airplane company which made fast, unstable, dangerous planes in the Thirties. The nearly cute name is a sham. They were killers. There were few pilots who flew Gee Bees and then lived to talk about it. Jimmy Doolittle was one. I was another.[22]

Just to add a little spice to the pre-race tension, there was a spat going on between the feisty Jackie Cochran and one of the other entrants, fellow-American Clyde Pangborn. He had been the original nominee to fly the Gee Bee, but a dispute with the aircraft's manufacturer, and a more lucrative offer from Warner Brothers to crew with Roscoe Turner in the Boeing airliner, meant that he had withdrawn. However, he was demanding money from the aviatrix for the transfer of his entry, and she was not budging. Even as she taxied in to the hangars and switched off, she was being asked by reporters about the dispute and all she would do was to hold up a card with the words: 'I have no comment to make. Please consult my lawyers.' She then sat in the cockpit for several minutes combing her hair and adjusting her make-up before climbing out. Her only comment on the dispute in later years was, 'If I'd been a man, we would have gone out behind the hangar to fight it out.' She went on in the post-war years to become the first woman to fly faster than sound.

The week before the race was a hectic one, with the de Havilland works team working day and night making last-minute adjustments and modifications to the three Comets. Sir Geoffrey de Havilland recorded in his autobiography many years later: 'I don't think that any of us would care to live through the weeks that preceded the start of the race.'[23] All the competitors' machines had to be scrutinised for airworthiness, be weighed in and have their fuel consumption checked, whilst pilots and their crews were test flying, preparing their equipment, poring over maps and generally discussing the intricacies of the route ahead.

The Mollisons with their all-black and gold-trimmed DH 88 Comet Racer, G-ACSP, at Mildenhall in October 1934. The name Black Magic *had yet to be painted on the nose cowling. Note the small discs mounted on the front of the spinners. These were part of the rather crude Ratier mechanism which was used for changing the propeller blades from fine to coarse pitch during flight.* (Times Newspapers)

A certain amount of friction arose between the Mollisons during this week, mainly because Jim would not agree to Amy flying the Comet on her own before the start of the race. She insisted on doing so, reminding him that she had paid 50 per cent of the machine's cost and was therefore fully entitled to fly it. There is no doubt her pride was at stake in front of others, since by denying her the chance to take the Comet up alone, he was openly showing that he had no confidence in her flying ability, or so she thought. Some say that Jim stormed off to increase the amount of insurance cover before downing a few swift brandies at the bar, as Amy made for the hangar to give instructions to the de Havilland ground staff.

'Please prepare *Black Magic* for take-off. I'm taking it up!'

It has to be remembered that Amy had never flown anything faster than the Beechcraft she now owned, and that the Comet was a tricky aeroplane to fly. It not only had a retractable undercarriage and flaps to think about, it also had a rather crude mechanical device for varying the pitch of the propellers. It used an inflatable football bladder in the propeller's spinner to hold the blades in the their fine setting for take-off. Once the aircraft was in the air and a forward speed of 160 m.p.h. was achieved, a small disc protruding from the nose of the spinner was forced back to release a Schrader valve, causing the bladder to deflate and allowing the blades to move into coarse pitch. This meant that one had a fine setting to give maximum power for take-off, and a coarse setting to give optimum cruising speed at higher altitudes. The device can be likened to a car's gearbox, where one selects a low gear for hill climbing and a higher gear when cruising on a motorway. However, the system used on the Comet had the disadvantage that the blades could only be reverted back to fine pitch by reinflating the bladder whilst the aircraft was back on the ground. This meant that the system could prove hazardous if a pilot suddenly needed to abort a landing and go round again – something Amy was prone to do. However, she handled the machine superbly and after thirty minutes in the air, she made as good a landing as any of the other Comet pilots.

Amy's parents were at the start of the race at 6 a.m. on Saturday, 20 October to wish their daughter and son-in-law good luck. They mingled with the raincoated crowds gathered around *Black Magic*, which was now in the line-up ready for take-off. There were five mandatory control stops along the route – Baghdad, Allahabad, Singapore, Darwin and Charleville – before they reached Melbourne, some 11,300 miles distant. The Mollisons had drawn first place in the ballot for starting position and they were to be followed at forty-five second intervals by the remaining competitors.

As the Mollisons were helped into the narrow confines of the aircraft's cabin, one of the observers on starting duty, Flying Officer Jeffrey Quill, recalled:

> I saw Jim and Amy climb into their Comet. He looked as white as a sheet and as if he had been sloshed for forty-eight hours – which I suspect he was. She looked very nervous and apprehensive and I felt desperately sorry for her having to climb in behind that raffish character for such a venture as this.[24]

The starter's flag fell at 6.30 a.m. precisely, and *Black Magic*'s two powerful Gipsy engines responded smoothly and eagerly to the throttles, much like dogs straining on the leash. Behind them the wet grass flattened out into a fan-shaped semicircle as the black Comet began to pull away. In the cabin the noise from the tailskid's scraping surrendered to the intermittent thumping of the undercarriage as the machine's wheels sped over the rough turf and the tail lifted. Carefully countering any tendency for the aircraft to swing, Jim could now see the far side of the aerodrome directly ahead of him. Within 650 yards they were airborne and turning onto a south-easterly heading. Ahead of them lay the longest leg of the race, with Baghdad 2350 miles away on a great circle route.

Meanwhile, Amy was making sure that they were on course for the Iraqi capital. Unlike the two American airliners, none of the Comets carried radio equipment, preferring to take extra fuel. This meant that Amy's navigation relied solely upon the dead reckoning method, calculating the course with an aperiodic compass, their known airspeed and estimated drift. Apart from these they relied upon sighting visual landmarks to corroborate their position on the strip-map. Arthur Clouston, destined to become the most experienced of all the Comet pilots, believed that it was best to ignore drift because of the unreliability of forecasting wind strength at various altitudes. He found that the variations generally cancelled each other out.

The wheels of *Black Magic* sent back a swirl of red dust as Jim and Amy landed at Baghdad at 7.10 p.m. Turning and taxiing towards the floodlit control point, they were met by an eager crowd of officials, who confirmed that they had covered the distance from Mildenhall at an incredibly fast average speed of 199.77 m.p.h. Their navigation had been spot on in spite of foul weather. As one of the works mechanics climbed up onto the wing to swing the canopy aside, the Mollisons' first words were, 'Have any of the others been here before us?' With the noise of the engines still ringing in their deafened ears, they received the reply they were longing to hear: 'No, you're the first.' Not only were they the first to arrive, they were the only ones to make the Iraqi capital non-stop. It was a superb piece of airmanship and navigation.

In the meantime, news was coming in from some of the less fortunate competitors. Jackie Cochran and Wesley Smith had retired at Bucharest. They were flying at 14,000 feet over the Carpathian mountains when their engine cut out within seconds of switching to their reserve tank. In the ensuing panic their first reaction was to bale out, but when Wesley saw that his partner's canopy was jammed and she could not get out, he chivalrously decided to stay with the machine. Jackie then discovered that the fuel switch had been incorrectly marked, and that 'on' in fact meant 'off'.

In spite of surviving the scare, they hardly needed any persuasion to put down at the Romanian capital. They were extremely cold and had been wearied to the point of exhaustion by the raucous noise from their big Hornet engine. Worse was to come, however, for when Jackie made her landing approach she found that one of the wing flaps was inoperative. She made two attempts at a landing, but the imbalance of the control forced her to go round again. On the third attempt, with one flap down, she hit the turf with all of the Gee Bee's customary clumsiness and needed the whole length of the airfield to pull up. The two Americans were relieved to retire with no more than a damaged undercarriage.

Others were far less fortunate. The two New Zealanders, Gilman and Baines, were killed when their Fairey Fox crashed in flames in southern Italy during an attempt at an unscheduled landing in strong winds, whilst Parer and Hemsworth in a similar machine retired with a leaking radiator. Captain Neville Stack in an Airspeed Viceroy got as far as Marseilles before retiring with brake problems, and Harold Brook in a Miles Falcon also pulled out of the race in France. Further along the route the Lockheed Vega flown by Jimmy Woods, with Don Bennett as his navigator,* retired when they suffered a heavy landing at Aleppo.

Shortly after the Mollisons left Baghdad to head for Karachi, the beam from the landing-lamp in the nose of Scott and Black's Comet *Grosvenor House* appeared as a pin-prick of light in the darkened sky. They were the second competing aircraft to land at the Iraqi control point. Minutes later the near empty tanks of their machine were being filled as the two men questioned

---

* The man who went on to become the legendary Air Vice-Marshal leading the RAF's wartime Pathfinder Force.

officials about the arrival and departure of *Black Magic*. So near were they to seeing their arch rivals that Scott was to comment later in his memoirs: 'There [Baghdad] they told us that the Mollisons had left fifteen minutes earlier *on a course for Jask*. The dust kicked up by their black Comet still overhung the aerodrome; she had shaken it from her dainty fabric with a vengeance.'[25]

By choosing to overfly Jask on the coast of the Gulf of Oman on their way to Karachi, and then to proceed to Allahabad, the Mollisons were conceding not only an extra distance of 180 miles, but also the time wasted in an additional stop. There could only be two possible reasons for their decision: either they thought that Scott and Black were out of the race completely, something they had good reason to believe since they had spent ninety-five minutes on the ground without sight of their main rival; or they were determined to add the UK–India record to their laurels. Maybe it was for both reasons. Charles Scott, however, knowing that the Mollisons were making for Karachi and not Allahabad, made a bold decision. He would do the 2300 miles to the Hindu holy city in one hop and take out his rivals' lead.

Exactly twenty-two hours and thirteen minutes since they had left Mildenhall, the Mollisons touched down at Karachi on the Sunday morning. They had set a new UK–India record by knocking twenty-eight hours off the previous one. Neither Amy nor Jim was a stranger to the city, and they were warmly received by the awaiting European and Indian crowd. Jim had served here in the RAF in the 1920s, and had landed at the aerodrome when he broke Scott's Australia–England record three years earlier, whilst Amy had been welcomed in the city when she arrived there in 1930 after knocking two days off Hinkler's time.

An hour later, the couple were heading for Allahabad when disaster struck. As soon as Jim tried to wind the undercarriage up, he knew that something was wrong, for the wheels would only partially retract. Fortunately, he found that they would go fully down and they were forced to return to Karachi. Mechanics were soon working frantically to rectify the fault, but it was another six hours before the problem was solved.

Even then their troubles were not over, for within two hours of departing for the second time they were back. History gives two different reasons. The first is that they had forgotten a vital map, and the second, that the maps were 'unsuitable'. All we do know is that a rather dejected Jim told reporters, 'Rather than carry on with suspect landing-gear and inadequate charts, we will spend the night here.' Both Amy and Jim realised that an immediate departure would have meant a night landing at Allahabad, and this they evidently were not prepared to risk. No doubt the memory of their mishap at Bridgeport in the *Seafarer* had unnerved them both. To add to their troubles, a heavy mist was now settling over the aerodrome, so they decided to go to a local hotel and snatch a few hours' sleep while they could.

At midnight local time, they were roused from their slumber for a return to the airport, where they were greeted with the devastating news that Scott and Black had left Allahabad twelve hours earlier; worse still, they were in all probability now flying down the Malayan coast to Singapore. Moreover, the American and Dutch airliners had arrived and departed from Karachi without a hitch. The Mollisons' only small crumb of comfort was in seeing Bernard Rubin's green Comet being worked on as they left. They had now lost sixteen precious hours.

Ahead of them in the dark that night was the radio-assisted Boeing airliner of Roscoe Turner and Clyde Pangborn, two experienced pilots who were having great difficulty in finding the Bamrauli aerodrome at Allahabad. Although it was a modern, well-equipped airfield using a powerful flashing beacon, which could normally be seen from a distance of 70 miles at night, the flashes of lightning from thunderstorms across central India confused the competitors. For Amy and Jim the bad weather was to have far graver consequences. After holding an easterly course for two hours, they noticed a discrepancy between the readings of their two compasses due to the electrical disturbances that they were encountering.

Such was the noise level from the two Gipsy engines that normal conversation from pilot to co-pilot was not possible, and tempers began to flare as a rapid exchange of notes passed between them. Adding to their difficulties was drift from a strong northerly wind of which they were unaware. Soon they were hopelessly lost in an all-enveloping Indian darkness. It was an inky blackness relieved only by the incandescent glow from the exhausts of the Comet's engines.

Sighting the lights of a town, they decided to circle above it until the dawn broke and there was sufficient daylight to risk a landing. As soon as they were able to pick a suitable spot, Jim put the Comet down safely in a field on the outskirts of what turned out to be the town of Jabalpur, some 175 miles to the south-east of Allahabad. By now relationships between husband and wife were at breaking point, and with little hope of retrieving the lost time, Jim threw all caution to the wind. Their fuel tanks were low and with little prospect of obtaining aviation fuel they took on commercial petrol from the local bus depot. At 10 a.m. they took off with throttles wide open, overdriving the aircraft at low altitude. Amy protested that this would only damage the engines, which it did.

As *Black Magic* was approaching Bamrauli, now on one engine, *Grosvenor House* was streaking along the coastline of Java on its way towards Darwin. Much to the bitter disappointment of the Mollisons, their arch-rivals were to take the first prize in the MacRobertson with a time of seventy hours and fifty-five minutes from start to finish. A crowd of 40,000 had gathered in the drizzling rain to greet Scott and Black as their red Comet appeared out of a grey sky to land at Melbourne's Flemington racecourse. Charles Scott's only comment on the race after he landed was, 'It was lousy – and that's praising it.' They were followed in by the KLM airliner twenty hours later, and in third place came the Boeing, less than three hours after the Dutchmen.

Upon their arrival at Allahabad the Mollisons' Comet was found to have damaged several of its engines' pistons and cylinder-heads through overheating, a problem caused by using low-octane fuel. The engine replacements they needed were not available and they were forced to withdraw from the race. Maybe the use of incorrect fuel was not the sole cause of their failure, since one of the aircraft inspectors on duty when they landed 'noticed three empty whisky bottles lying by the side of Jim Mollison's seat and convenient for his right hand'.

Perhaps the most poignant observation of all was made by Jehangir Tata, founder and chairman of Air India, who was at the airfield on the day, acting as a marshal.

> I was present when James Mollison and his co-pilot wife Amy arrived after some hours' delay. After landing, they remained for quite a while in the middle of the airfield, whereupon an Indian Air Force NCO was sent in a jeep [sic] to enquire what the trouble was and to help them. He brought them in, leaving their Comet behind to be towed in later. He told us that when he had gone to them on the airfield he found them having a heated and loud discussion, or quarrel, as to who was responsible for their engine trouble, later found to be due to their having refuelled at Jabalpur with ordinary petrol.
>
> Amy left the same day, but Mollison remained in the reception tent for most of the time during the next two days, unchanged and unshaven, not having brought even a change of shirt with him. As was to be expected, he was not in a very happy or communicative mood and I, naturally, didn't inflict myself on him.[26]

By now it must have been all too obvious to both Amy and Jim that not only was their flying partnership a disaster but, more importantly, their marriage was unravelling.

Chapter Seventeen

THE FRENCH HOTELIER

few days after their arrival at Allahabad tempers subsided sufficiently for Jim to join Amy in Calcutta for two weeks whilst spares were flown out from the de Havilland factory and repairs carried out on the Comet's damaged engines. The Mollisons then flew their aircraft back as far as Cairo and booked into the Heliopolis Hotel, enabling them to make a sight-seeing tour of the city. Their relationship must have remained cool because whilst they were in the Egyptian capital Amy decided she had had enough of the cramped confines of the Comet, complaining that there was 'barely enough room to carry a toothbrush'. She caught the KLM flight from Cairo to Amsterdam and then flew to Croydon, leaving Jim to fly their plane home.

Amy had now come to the firm conclusion that she must forge her own career, separate from her husband's. The door of opportunity with Hillman Airways had closed on her with the death of its founder, and her contract with the *Daily Mail* had not been renewed. Admittedly, she still held the UK agency for Beechcraft and made £200 on each of the few aircraft she sold, but there was still no regular income and she was obliged to eat into her fast-dwindling capital. However, she was never short of ideas, some wildly impractical, such as appearing on the music-hall stage, or writing air fiction in collaboration with a known author. A more realistic venture she envisaged was to start an air-tour company to fly the rich and famous on holiday packages to luxury hotels abroad. In this respect she was thirty years ahead of her time, for the charter flights that we take for granted today were then but a distant dream.

Jim, who had now told Amy that he preferred to lead a bachelor-type existence, was willing to keep up with the pretence of a happy marriage as long as it allowed them to reap the advantages of their celebrity value. They decided to sell *Black Magic* to the highest bidder, and were pleasantly surprised to find that its value had risen considerably above what they had paid for it. Prospective buyers were impressed by the fact that the three Comets had, between them, scored an outright win in the MacRobertson, set a UK–India record, and gained a fourth place in the race. As a result, the three aeroplanes were in strong demand, not only from the Air Ministry, which eventually paid A. O. Edwards £7000 for *Grosvenor House* – a 40 per cent profit on what he had paid for it – but also from foreign buyers. *Black Magic* was flown to Hatfield for a complete overhaul before being sold to the Portuguese government in February 1935. To the last it retained its black and gold livery, only now it was renamed *Salazar* after that country's prime minister.

Apart from participating together in an aviation debate organised by the Women's Engineering Society early in January (Amy had been elected president of the society for the period 1934–7), the couple saw little of each other until the summer. Without any word to his wife, Jim sailed for the USA in late January after being interviewed by the *Daily Telegraph*, which reported that he was 'about to attempt a stratosphere flight'. It was in fact a feeble ploy to give some cover of respectability to their estrangement over the next four months. Amy was desperately unhappy, and spent a brief riding holiday in Malvern with her youngest sister, Betty, before sailing to Madeira in search of some sunshine.

Jim arrived in New York in the depth of one of that city's harsher winters to link up with some of the social contacts he had made previously. Since his 1933 visit the Prohibition laws had been repealed and with them had gone much of the corruption that had dominated so many of America's big cities. However, this did little to diminish his fascination with crime, something which seemed to excite the prurient streak in his nature. This comes out quite forcibly in *Playboy of the Air*, in which he described being privy to police raids on Harlem drug parties in New York's 43rd Precinct, reputedly one of the toughest in that city. Through a police contact he was 'invited' to attend identification parades where the unrestrained use of the rubber truncheon was quite commonplace.

After spending a short time in New York, he left Manhattan in snow and ice to step from an airliner at Miami into a world of palm trees and white sands that compared with the best that he had seen at Bondi. All his predatory instincts as a womaniser were revealed in his book as he eulogised the tanned, high-heeled beauties on Collins Avenue and Palm

Beach. In the latter haven, he was welcomed by the affluent members of the '400 Club', an elitist group of individuals whom he described as being as exclusive as any amongst the British aristocracy and where, he boasted that the rich and famous vied for his patronage as their guest. There is a strange irony in the fact that he received more acclaim and recognition in the USA during his lifetime than he ever did in the UK. Maybe the fact that two airports in North America were named after him in the 1930s goes some way to explaining why.

One particular facet of American society that fascinated him was the difficulty of identifying the social position and status of a person by their accent. In Britain it presented little problem; one knew instantly by the voice alone whether one was speaking to a shop-girl or to a baroness, but not so in the States. On one occasion, when the hotel's female receptionist purred over the room-service telephone, 'Yes, Captain Mollison, can I help you?', he was seduced and fooled by the honeyed tones of someone he later considered to be his social inferior. The nuance in her voice was not to be missed and the invitation was quickly pursued. Small wonder there was outrage when these encounters were described in his book. One can imagine how incensed Amy must have been when she read them. It is doubtful if she ever forgave him for embarrassing and humiliating her with these 'kiss-and-tell' revelations.

Amy returned from Madeira in April 1935 to suffer a series of misfortunes. After staying at St John's Wood with some friends whom she had met on holiday, she drove to her parents' home in Bridlington. Whilst she was visiting them she was involved in a road accident in which a man was killed. She was driving her Mercedes on the Fraisthorpe by-pass when an approaching motorcyclist pulled out from behind a car he was attempting to overtake. As he did so, he clipped the car's rear bumper and was thrown directly into the path of her car, with the result that his pillion passenger received fatal injuries. Although Amy was exonerated of all blame at the inquest she was required to attend, it must have unnerved her, for not long after she suffered a nasty accident when she failed to lower the undercarriage on her Beechcraft cabin-biplane whilst making a landing. She attempted to make light of the matter in a letter to her father, but could not conceal that it had shaken her confidence.

Upon her return to London she moved into to the Savoy Hotel, where her celebrity status appears to have been valued by the management much as it had been before at the Grosvenor. However, in spite of the luxury, she was not happy separated from her husband and she implored him when he returned from the States a few weeks later to come back and live with her.

He refused outright, saying that he preferred to continue with his single lifestyle. He always maintained that he could do little right in his wife's eyes, and that she was too demanding and quick-tempered.

However, whether out of sympathy for her two accidents or not, in July he relented and agreed to go back and live with her. It was a short-lived reunion, for one evening whilst Amy was at the theatre he invited one of his new conquests back to the apartment. Amy returned unexpectedly early and found the two of them lying on the bed in her bedroom. Whilst they were not *in flagrante delicto*, a blistering scene followed in which Amy told the woman to get out. Jim pleaded that nothing had happened between them and that they were merely sleeping off the effects of too much drink. In the ensuing row Amy ordered her husband to leave immediately, but he refused, still protesting that he had not been unfaithful to her. With her suspicions now fully aroused, Amy left the hotel early the next morning and went to stay with a friend, leaving Jim to his own devices.

Whilst the Mollisons maintained an outwardly polite relationship over the next few months, they lived apart and there was never a true reconciliation. Amy was now more determined than ever to go her own way, and during this period her flying log shows that she began to make quite lengthy trips in her Beechcraft to Paris, Biarritz, Vienna and Budapest. In one of the entries the name 'Diana' appears, no doubt referring to the legendary Diana Caldwell, who was at that time in a whirl of social activity, mixing with the aristocracy of Europe. Earlier she had run a small cocktail club called the Blue Goose in Bruton Mews, just off Bond Street, which was probably frequented by the Mollisons. Famed for her *femme fatale* persona and classic good looks, she was to become the woman most heavily involved in the infamous Lord Erroll murder trial in 1941.

James Fox, in his compelling account of Kenya's Happy Valley set in his book *White Mischief* – later to become a Hollywood movie – throws some light on the kind of circles that Amy was now moving in. He wrote:

> She [Diana Caldwell] led a hectic social life, and used her aeroplane to pursue it across Europe. She flew with the famous aviatrix, Amy Mollison. She fell in with an exclusive group of aviating aristocrats, one of whom was Prince Starhemberg, the Austrian Vice-Chancellor. They would meet in Budapest and the *Tatler* correspondent was somehow always there . . .[27]

Pictures taken during this period show Amy on board a river-launch with a rather youthful-looking Randolph Churchill and Claire Luce, the woman with whom he was currently enjoying a flirtatious affair.

Amy's Beechcraft D17 – L, G-ADDH, at Castle Bromwich on 23 February 1936.
(E.J. Riding)

Prince Ernst Starhemberg was a playboy and right-wing political adventurer who had originally sympathised with Hitler and his anti-Semitic views in the early twenties, but subsequently became disenchanted with the German dictator when he revealed his plans for annexing Austria to the Third Reich. Starhemberg, himself a keen pilot, would have known Amy through his interest in aviation and through her visits to Vienna, not only in 1930 whilst making her solo flight to Australia, but also subsequently. It was most certainly around this time that she also first met the tall, urbane François Dupré, a wealthy Parisian hotelier and owner of a stud-farm for the breeding of racehorses.

François Louis Jules Dupré came from a wealthy background and was the grandson of a famous French painter. He originally set out to become a financier and joined the staff of a Paris bank, from where he was sent to train as a manager at Barclays Bank in London in 1912. Whilst working in London he became very friendly with Melville Greenshields, who ran the City office of Greenshields Partners. As a result of this friendship, Dupré was offered a more lucrative post with Greenshields and was subsequently appointed manager of their Paris office, where he remained throughout the First World War.

Dupré's astuteness was not confined solely to banking and finance; it also extended to his ability to marry into the right kind of family. This he did most successfully when he married the Duke of Polignac's widow, Daisy

Singer, one of the daughters of the Singer sewing machine magnate. With the wealth of the Singer family behind him and a zeal for hard work, he was able to become the major shareholder of the Hotel George V, the Plaza Athénée and La Trémoille in Paris. His cold and aloof manner earned him the nickname Le Baron from those who worked for him. One such was his former minute secretary, who recalled that he was 'a very forceful, shrewd, knowledgeable, tough character who knew what he wanted in life, particularly from hotels, horseracing and females – what a combination from a robust Parisian!'[28]

When François Dupré first met Amy, he was quickly attracted to a woman of such international fame and they formed what was for her a platonic relationship. He was shrewd enough to realise that his romantic notions towards her were not being reciprocated, but at the same time he was prepared to play a long, waiting game in the hope that if he himself did not attract, then his considerable wealth might. With a woman's intuition she sensed the direction in which he was moving the pieces and was prepared to go along with the game. When she told him of her ambition to form an Air Cruises company, which could fly his wealthy hotel guests to exotic holiday destinations, he became interested and responded to her idea by putting up £5000 for the purchase of a brand new de Havilland Dragon, to be fitted with luxury seating and radio equipment. Moreover, he invited her to be his guest whenever she happened to be in Paris by making a suite of rooms available to her at the George V Hotel.

Restless as ever, Amy rented a furnished residence called Belgrave Cottage near Eaton Square as her London base in November, and here Jim visited her before sailing for New York on the *Champlain*. During their discussion, Amy suggested that they should divorce, but Jim's answer was that they should wait a further six months until he returned from his round-the-world trip. He obviously did not want to shut the door on the possibility of making a further money-making flight with her. He took the line that although the passion had gone, the partnership need not. Probably with one eye on her public image, Amy agreed to wait.

In spite of the added financial security that Amy now enjoyed from her French benefactor, she could not resist the lure of the travel and adventure that long-distance record-breaking flights brought. Throughout the year she had read reports of Amelia Earhart's sensational 2400 mile solo flight across the Pacific from Honolulu to Oakland in California in January 1935; and then of Jean Batten's record time of seventeen days and sixteen hours from Australia to England in April, which meant that she became the first woman to fly solo in both directions (she had already broken Amy's

England–Australia record by over four days in May 1934). The busy little New Zealander then proceeded in November to make a solo flight across the South Atlantic in a Percival Gull, in which she knocked four hours off Jim Mollison's time.

Amy became more and more conscious that her name was slipping from the headlines, and when Bill Courtenay approached her with the suggestion that she made a further attempt on the Cape record, she needed little persuading. The amiable Tommy Rose, a good friend of the Mollisons, had recently set a new record time with a flight to Cape Town and back in a Miles Falcon monoplane. What was most striking about his flight was the cool, matter-of-fact way in which he made it. Press pictures taken at the time show him stepping out from the cabin onto the wing of his aircraft dressed in a lounge suit, and with his trilby hat in hand, looking very much like a businessman stepping onto the platform of a London mainline station on his way to the office. It was no longer the era of the leather helmet and flying goggles. This was the understated, fashionable, eye-catching way in which Courtenay advised Amy to conduct her flight.

Courtenay's suggestion seemed to fit in with Amy's idea of starting Air Cruises, a company which Dupré was prepared to launch in May 1936. A successful flight to the Cape and back in April would, she thought, provide worthwhile publicity for the new company. The problem of finding a suitable aircraft was partially solved when one of her many admirers, Captain H.L. Farquhar, offered her the use of his Beechcraft, which was fitted with a much more powerful engine than her own machine and therefore capable of setting a new record. On 5 March she flew the biplane from Heston to Castle Bromwich, but once more had difficulty mastering the retractable undercarriage, and as a consequence damaged the machine on landing. Fortunately, when Edgar Percival heard of her intention of taking a crack at the Cape record, he offered her the loan of one of his Percival Gull-Sixes. His sole motive for doing so was that it might promote his company's product, in much the same way as Jean Batten's successful flight to Brazil had done.

At this time François Dupré was in Canada, where he was pursuing a business interest in the Ritz-Carlton hotel in Montreal. Amy sent a cable informing him of her plans for the proposed Cape flight, as she was not sure how he would react to the idea now that the Air Cruises company was about to start operating. More importantly, it meant that her role as chief pilot would have to be delegated temporarily to someone else. Her cablegram gave him the opportunity to reveal his true feelings for her openly by wiring back that he was missing her, and that now they were 'business partners' they should spend more time together. It was not exactly the kind of message she wanted to hear, since she was not attracted to him romantically whatsoever. However, he was enthusiastic about her flight, had every confidence that she would succeed, and signed off by saying that she was constantly in his thoughts. This must have left her slightly uneasy.

The sun glistened on the burnished aluminium surface of the American Airlines' Douglas DC2 as it slipped in over the San Gabriel mountains and touched down on the oil-streaked runway of Glendale's Grand Central airport. Jim Mollison's first glimpse of Los Angeles and its lifestyle came as he entered the main lobby of the imposing, faux-Spanish terminal. Its architecture dominated the skyline and was possibly already vaguely familiar to him as a backdrop to so many of Hollywood's early movies.

Ever since the introduction of the talkies, there had been no bigger magnet for aspiring young actors and actresses than this particular part of Los Angeles. It was a place where supply outran demand and where these hopefuls ended up either as petrol pump attendants or waiting at table, in between standing in queues outside the gates of some film studio in the hope of gaining a bit part in a 'B' movie production. As soon as the Scot entered the airport's restaurant, his head was swivelling with all the consummate skill of a studio's talent scout. After being shown to his table by a slinky hostess, a Constance Bennett look-alike, and waited on by Hedy Lamaar's double, it began to dawn on him that this was the unbelievable norm of the San Fernando valley. What temptation for a man estranged from his wife!

Amelia Earhart had originally invited the Mollisons to be her guests at her Toluca Lake, North Hollywood home, and to attend an aero exhibition that was taking place in Los Angeles in November 1935. Understandably, with the marital rift between them, Amy had declined the invitation. As a consequence, Jim spent the next eight weeks making the most of his reputation as the playboy flyer. Through the Putnams he was given an introduction to the film studios, where he met many of the stars and was invited back to their homes. There is no doubt that he was welcomed for two reasons: firstly, the Americans loved a pukka, 'Dammit, sir', English accent – a whole colony of British actors and actresses were there already in Hollywood to prove it, and secondly, whilst the film industry never openly courted notoriety, it certainly attracted it. Jim Mollison qualified on both counts.

On 23 January 1936 the *Los Angeles Times* featured a picture of 'Colonel'

James A. Mollison alongside an article giving the aviator's views on the accession of King Edward VIII as the new monarch of half a billion subjects of Great Britain and the Commonwealth. There was considerable interest in the USA because of his much-publicised liaison with the American divorcée, Wallis Simpson. It was too good an opportunity for Jim to miss and he was able to share some of the personal anecdotes of his meeting four years earlier with the man who was then the Prince of Wales.

A week later the same newspaper carried a much less flattering report, about the aviator being charged in the Beverly Hills area for drunken driving and taken into custody overnight. He was released the next morning when a friend paid the $10 fine on his behalf.

It was all very much par for the course.

Amy's first attempt on the Cape record in April 1936 ended in near disaster when her pale turquoise Percival Gull ground-looped as she began her take-off run across the stony surface of Colomb Bechar's aerodrome on the edge of the Sahara. She was just about to start the second stage of her outward flight, when the aircraft skidded and slewed around in a cloud of dust. The starboard wing scraped the ground and the undercarriage was ripped off. With enough fuel on board for a 2000-mile hop, she was fortunate to escape without a fire erupting. Uninjured, she just sat in the cockpit and wept.

The Percival Gull Six, G-ADZO, at Colomb Bechar after Amy ground-looped whilst taking off for the Sahara desert crossing in April 1936. (Alex Henshaw)

In a letter to her father written six days later, she blamed a burst tail-wheel tyre as the cause of the accident, but it might well have been that she failed to take corrective action quickly enough as the aircraft met with a gusting side wind. A lesser mortal might well have given up there and then, but quitting was not a word to be found in Amy's vocabulary. With the help of the French authorities on the aerodrome and assistance from Peter Reiss, the damaged machine was repaired and flown back to the Percival works at Gravesend. She would try again.

When Jim Mollison returned from Australia, he must have wondered just what kind of reception he would receive as he rang the doorbell of Amy's newly rented Gate House residence in Ennismore Gardens, Kensington. In view of the imminence of her departure for South Africa, it is unlikely that there was any enthusiasm for scathing recriminations on her part. They agreed on a temporary truce until she returned, whereupon they would then make a serious reappraisal of their marriage. Amy certainly appears to have accepted the situation with more than a little good grace, since she is on record as telling her parents that her husband would look after her affairs whilst she was away.

Dressed in navy blue flying clothes designed specifically for her by Schiaparelli, Amy left Gravesend on the morning of Monday, 4 May to try again for a record flight to the Cape by the west coast route. This time, she not only beat Tommy Rose's time for the outward flight by eleven hours but, with only a three day break as against his three weeks, she set a new record for the return journey. It was a remarkable achievement by any standard, since not only did she fly home by what was for her an unknown route, via Johannesberg, Khartoum, Cairo and Athens, but overall she knocked one and a quarter days off her rival's time for the 14,400-mile flight. There was only one really unnerving moment and that was again during a landing. By the time she reached Cairo, she was on the point of exhaustion and she overshot the aerodrome to finish up perilously close to the boundary fence.

Whilst Amy was in Cape Town, Jim Mollison announced that he and his wife intended to make a round-the-world flight at equatorial latitude sometime in the near future. Whether he had discussed this plan with Amy beforehand is doubtful, and it might well have been that he was bouncing her into it. Amelia Earhart had been toying with the idea of making a similar flight as far back as November 1935, and no doubt Jim had picked up the rumour of it whilst he was in Los Angeles staying with the Putnams. By July the following year it was no longer a rumour; the American aviatrix took delivery of a twin-engined Lockheed Electra capable of flying 4500 miles non-stop. It was not until July 1937 that the flight actually took place and she and her navigator, Fred Noonan, lost their lives.

There was a certain amount of unpremeditated reconciliation going on between the Mollisons whilst Amy was away. As soon as she landed in Cape Town she was on the telephone to Jim, and later told reporters, 'I feel very lonely without him.' He for his part wired her back several messages before she left for home, one reading: 'Take no undue risks. I don't want to be a widower just yet. Jim.' Whether these exchanges were sincere or not it is hard to tell, although one feels that on her part they most probably were.

Jim Mollison bids his wife farewell as she prepares to depart from Gravesend for the Cape in her Percival Gull Six, G-ADZO. (T. Davis/Ron Neudegg Collection)

Amy waves from the cabin of her Perival Gull Six, G-ADZO, as she prepares to leave Gravesend on her record-breaking flight to the Cape and back on Monday, 4 May 1936. (T. Davis/Ron Neudegg Collection)

Gravesend Airport from the west in 1935. The dual control tower and clubhouse is on the left, then the airport's offices and customs post. The two smaller hangars belonged to the Percival Aircraft Company and Gravesend School of Flying. (T. Davis/Ron Neudegg Collection)

Back in London, where again she received a tumultuous welcome from the crowds, she was forgiving when her husband confessed to her that he had been unfaithful on many occasions, once within just a month of their wedding. They managed to patch things up sufficiently to attend various functions around the country together, notably the formal opening of Sewerby Park, near Bridlington, on 1 June, where today her many trophies, awards and momentos are kept as the Amy Johnson Collection in the art gallery of the Hall. It says much for Amy's unwavering popularity that even at a minor event such as this one, 15,000 people turned up to see her.

Spending time with Jim in this way meant that she had to forgo the part that she would have wished to play in her Air Cruises venture. Dupré could only accept the situation, and was forced to look for another pilot. Having an eye for the ladies, he chose the aviatrix Beryl Markham. She had been introduced to Dupré by her estranged husband Mansfield Markham, who was a very good friend of the hotelier.

Beryl was acknowledged to be the archetypal enchantress and had quickly learned that doors readily swing wide open for pretty blondes with charm. This is most clearly illustrated by the fact that she even cajoled her ex-lover and mentor, Tom Campbell Black, who was at that time recently married to the cabaret artiste Florence Desmond, into giving her practical flying tuition in Dupré's luxuriously equipped de Havilland Dragon. Although she had already logged some 2000 hours over a six-year period in Africa as a game spotter, they were all accrued on single-engined planes. She needed the tuition in order to get her licence rating for twin-engined aircraft. Amy must have wondered if she would ever regain her position with Dupré, even though he was still besotted with her and urging her to divorce Jim.

By June 1936 the Mollisons had dropped the idea of a round-the-world flight when they failed to buy back the Comet they had sold to the Portuguese. However, in the same month, the wealthy South African industrialist, I. W. Schlesinger, announced that he was prepared to put up £10,000 prize money to sponsor an air race from the UK to South Africa, which would be run on similar lines to the MacRobertson. Its aim was to celebrate the opening of the Johannesberg Empire Exhibition, which was due to commence in September, and at the same time give the British aircraft industry a chance to display its wares. It was an ill-timed event, for Germany was beginning to flex its military muscle by occupying the Rhineland earlier in the year, and as a consequence the rearmament of the RAF had now become a priority. Furthermore, the fact that entries had been restricted to those flying British machines meant that there were eventually only nine competitors.

Jim lost interest in the Schlesinger when he realised that only the three extremely fast, single-seat Percival Mew Gulls had any real chance of success, and that they had already been allocated to various pilots, one of whom was Tom Campbell Black. Amy would only consider competing on her own since the *Black Magic* fiasco in the MacRobertson, and therefore opted out for the same reason as her husband. However, she was given a chance to fly one of the fast Mew Gulls in a quite unexpected way when Campbell Black was tragically killed at Speke aerodrome two weeks before the race. He was taxiing the *Miss Liverpool* (so named because he was being sponsored by John Moores, owner of the Liverpool football-pools syndicate), when he was struck by an incoming Hawker Hart bomber whilst looking down in the cockpit at one of his maps. The bomber's propeller sliced the Mew Gull almost in half midway along the fuselage, in the area of the cockpit. Black was rushed to hospital, but succumbed to his injuries shortly after.

Ten days before the start of the race, Amy announced that she was prepared to compete in the *Miss Liverpool* if Percival could repair the damaged machine in time. In spite of her protests that her intention in flying the machine was solely in honour of Tom Campbell Black, she was bitterly criticised. Many thought that she was just cashing in on Florence Desmond's grief in order to gain publicity.

As it transpired, her proposal did not materialise because the workshops at Gravesend were unable to complete the work in time. Jim kept his eye on the sleek little racer and purchased it two years later, whereupon, with a fair degree of nostalgia for his old ANA flying days with Kingsford Smith, he renamed it *Southern Cloud*. He even offered it to Amy to use in 1938, along with all his maps and equipment, when she was going through a bad patch, and so that she could boost her flagging finances by making a long-distance flight in it. By then she had lost all zest for record breaking and she declined his offer. With the outbreak of the Second World War the ill-fated machine was put into storage at Lympne, where it was destroyed by fire when the airfield was attacked by Stukas during the late summer of 1940.

Without any involvement in competitive flying, the Mollisons were thrown back more and more into each other's company and the strain on their marriage began to take its toll. Much as they tried to keep up the pretence of a happy marriage, from time to time the image began to slip. One such occasion was when they booked into the Norbreck Hydro in Blackpool for a brief stay that summer. One of the residents at the hotel recalled:

> In those days the Norbreck was teetotal and the nearest place that one could get a drink was across the fields at the Red Lion on the edge of Bispham village. During the night, Jim Mollison returned and there was a blazing row. Jim and Amy were arguing outside, which woke many of us up. Jim was unsteady on his feet and fell back into the ornamental pond outside the entrance, with the result that Amy became hysterical and was shouting about going somewhere. She left in a car and apparently went to the airport at Stanley Park, where she attempted to fly off in her aeroplane. Fortunately, she failed to do so. Someone there was able to persuade her that she was in no fit state.[29]

In one last desperate bid to save her marriage, Amy decided to spend the whole of August on holiday with Jim at Juan-les-Pins, on the Riviera. As far as she was concerned it was his last chance to show a change in his ways, but as the adage goes, the leopard cannot change its spots. Whilst we have no record of his behaviour on holiday, it was evidently sufficiently bad for Amy to approach her solicitors upon her return home, and to instruct them to commence divorce proceedings. There is every possibility that there was an undisguised flirtation going on at this time between Jim and the woman who was soon to become the second Mrs Mollison.

Amy asked Jim to leave their Gate House residence and he moved into the Hyde Park Hotel. Before long, he was being seen and photographed by the press dining and dancing with Beryl Markham in London's West End. She had left Dupré's employ with Air Cruises in an attempt to become the first

Beryl Markham, the first woman to solo the North Atlantic from east to west.
(Author's Collection)

woman to solo the North Atlantic from east to west. On the surface, Jim was merely advising her on the preparation she would need for such a flight, but she was known to be a man-eater and gossip soon became rife in Fleet Street.

Beryl had managed to find a sponsor for her proposed flight through an offer made by the wealthy Kenyan planter and keen amateur pilot, Lord Carberry. The two had known each other in Nairobi, and during a dinner party in a London hotel, Carberry offered to loan Beryl the Percival Vega Gull that he had ordered for the Schlesinger Air Race. He was an unpleasant character, a man who was known for his cruelty to animals, and some suggested that there was even a faint streak of malevolence in his apparent generosity to Beryl, since he stipulated that she must fly non-stop from England to New York, and return his aircraft before the start of the race. Not for one moment did he imagine she would accept his challenge and risk her pretty neck on an Atlantic crossing alone, but she did. Without flinching, she replied, 'You're on!'

With Jim Mollison at the controls of a hired Puss Moth and Beryl seated behind him, the couple flew into the airfield at Abingdon on 4 September under a storm-laden sky. Mollison did his best to persuade her to delay her flight, but to no avail. Just before she climbed up onto the wing of her aircraft in the late afternoon, he unstrapped his gold watch, the one he had used on his own Atlantic crossing in 1932, and handed it to her with the wry comment, 'Here Beryl, don't get this one wet.' Twenty hours later she upended her machine in a peat bog at Baleine Cove, Cape Breton, just 100 yards from the shoreline. Not only was she the first woman to fly that vast expanse of ocean alone in that direction, but she had bettered Jim Mollison's time by ten hours. Lord Carberry's jaw must have dropped just a little bit when he heard the news.

Amy wasted no time after she separated from her husband, immediately returning to Paris in order to resume the role she had vacated with Air Cruises. As soon as François Dupré discovered that there was an impending divorce between the Mollisons, he became even more persistent in his attempts to persuade Amy to marry him. He told her that he was madly in love with her and that he wanted her as his wife as soon as she was free. In spite of her forthrightness in telling him that she would never marry him because she did not love him, he acted as if he knew that given time she would change her mind. Nevertheless, it says much for her moral integrity that she was not prepared to be seduced by the riches and lifestyle that he so tantalisingly dangled before her.

Amy and François Dupré at St Moritz, 1936. (Amy Johnson Estate)

In a letter written from Lausanne, where she was staying with Dupré after having flown him to Switzerland in late September, she confided to her father the problems that she was still having with Jim.

> It was a great shock to find that he had gone to America. I didn't know and only saw it in the newspaper whilst I was in Paris. He didn't keep his promise to see me and settle things before he left. If he happened to go down in the Atlantic I suppose I get left with all his debts, as he has put every penny he possesses into his plane. He owes about £6000 as he has left without finishing off his Income Tax appeal . . . He was fighting it of course but the date for doing so has passed.

Unbeknown to Amy, Jim had negotiated the purchase of a fast and powerful two-seater monoplane from the Bellanca factory in Delaware, in which he proposed to make a record-breaking crossing of the North Atlantic, and then fly on to Cape Town. When Amy discovered which type of aircraft he proposed to use she was furious. In a letter to her parents, she wrote:

> I've just found out that the plane he's bought was one I wrote to New York for particulars of ages ago and never received a reply. The reply came but Jim opened the letter, kept it, said nothing to me and eventually bought the plane.
>
> Another thing I've just discovered today is that the last will I made in his favour is missing. He must have taken it, thinking it would be safer in his possession . . . Now of course I shall make a new one with Crockers.

Relations became even worse when Jim named his Bellanca aeroplane *The Dorothy* in honour of Dorothy Ward whilst he was in the USA preparing to make his Atlantic crossing. Ten days before he left Harbour Grace, Newfoundland, to set a record time of just over thirteen hours to Croydon, Amy was involved in an accident that resulted in a fractured nose and a dislocated shoulder. She was returning to Croydon from Le Bourget in her Beechcraft when she ran into dense fog over southern England. In her desperate bid to put down for an emergency landing at Chelsfield, near Orpington, the aircraft somersaulted onto its back and she was left dangling from the straps. Again she was fortunate to escape with such minor injuries.

After leaving the local hospital, where she was taken for X-rays and treatment, she was driven back to her home by a news reporter who had been quick off the mark in an attempt to scoop a story. As they pulled up outside her Gate House residence, they were met by a cluster of photographers and reporters who had been waiting for her to appear, having been told by the housekeeper, Jessamine James, to keep their distance on the other side of the front door.

Amy knew that if she refused to allow them to interview her about her relationship with Jim, it might only make matters worse. Sitting in an armchair with her arm in a sling, plaster on her nose and blood spots on her

Jim Mollison upon his arrival at Croydon Airport on 30 October 1936 in the Bellanca Flash, 28-70, The Dorothy, *after setting a new record for the North Atlantic crossing.* (Philip Jarrett)

hair, she parried their questions as best she could. They were not interested in the crash, only in the rumour that she and her husband were about to split up. She avoided any direct reference to a divorce, in spite of being offered £50 by the *Daily Express* for an exclusive story on the state of her marriage. Her statement was brief and to the point. 'I have decided to separate and live entirely apart from my husband. From now on I wish to be known by my maiden name.' A chorus of response came as one voice: 'Divorce?' To which she replied: 'I have no further comment to make. Good evening, gentlemen.'

The fact that Amy had announced that she was reverting to her maiden name, which was as good as confirming a divorce, prompted the *Daily Express* to tell its readers the next morning that she had just thrown £50 away. When the news reached New York it happened to coincide with Jim Mollison's arrival at Floyd Bennett Field where, apart from the publicity which he welcomed, he was given a great deal of sympathy. Amy was now seen as kicking her husband in the teeth at the very moment he was about to risk his life on a dangerous flight. If he did not survive the crossing and she was subsequently seen to be mourning him, she would then be seen as a hypocrite. If she showed no public grief, then she would be slated as being hard-bitten and uncaring. Poor Amy had placed herself in a no-win situation.

Amy spent the next few weeks with friends in Buckinghamshire, and whilst she was there recuperating, Dupré came to see her. He not only promised to take her for an exotic holiday over Christmas and New Year, when he would be visiting Morocco and Kenya to join Lord and Lady Furness on a safari expedition, but also, at her suggestion, to replace the obsolescent Dragon with a modern twin-engined Caudron Goeland passenger plane. The holiday never materialised owing to a breakdown in Amy's health in December, and she spent the Christmas period at a Swiss clinic with her sister Mollie as her companion.

Amy's poor health after her crash at Chelsfield and the war clouds that were now threatening to engulf the whole of Europe, caused Dupré to question the viability of continuing with Air Cruises. Reluctantly, the company was wound up in January 1937. However, he went ahead with Amy's suggestion to replace the Dragon with a Caudron passenger plane and assured her that she could use it whenever she wished. Moreover, in order to make up for the disappointment of the cancelled African safari trip, he accompanied her on a skiing holiday to Austria with one of his colleagues after she left the Swiss clinic.

Amy was never quite able to get the bug of record-chasing out of her system, and in 1937 she still clung to the idea of encircling the globe, much as Amelia Earhart was currently planning to do. Such a flight would cover a distance of some 25,000 miles in several stages, and it would need expert navigation crossing the vast expanse of the Pacific ocean. She had high hopes that Dupré would finance the venture by providing the machine, but in addition she needed someone who was expert in celestial navigation to crew with her.

At the end of March she sailed for New York, where she met up with Dupré, who was on one of his many business trips to Canada and the USA. Amy was quite used to the Frenchman showing her pictures of himself with some of his glamorous girlfriends in Budapest or Vienna, no doubt to remind her that she was not the only one in his address book. Even if this did not tempt her into entering a marriage of convenience, it no doubt reminded her that she was an unattached woman fast approaching middle-age.

One of the reasons for her New York visit was that she was dissatisfied with her Paris dentist and had been recommended to a Manhattan orthodontist who could cap and crown teeth to the perfection required by Hollywood. Another was that she had become friendly with the well-known beautician Gloria Bristol, who ran a high-class salon on Fifth Avenue. They had met in France and found that they had much in common, for the American also flew her own private Beechcraft. Portraits taken just after her US visit show Amy elegantly coiffed and impeccably made up, the epitome of sophistication.

The main reason for her American visit was to improve her flying skills, and in order to do so she had booked a fortnight's course with a school of navigation run by Commander Philip Weems, a man renowned as one of the world's greatest authorities on the subject. The modest retired US naval officer ran his own private school from his Annapolis home. He numbered amongst his pupils some of the most illustrious names in the history of aviation, including pioneers such as Lincoln Ellsworth, Captain A. C. Read, Charles and Anne Lindbergh, Fred Noonan and Harold Gatty, one of the finest aerial navigators.

Whilst she was with Weems Amy was taught how to use the bubble-sextant, an instrument which would enable her to gain a fix on her position from the sun or the stars, even when she was above cloud and unable to sight the horizon. This instrument, with the aid of accurate chronometers, reference tables and a map, enabled an aerial navigator to determine an aircraft's precise position. Weems claimed that Pan American Airways' planes were currently able to pick up an island 200 feet wide and 4 feet high with the method he taught. It was essential training for someone making a round-the-world flight by way of the Pacific, which involved a 7000-mile ocean crossing.

Amy realised, much as Amelia Earhart had done before her, that it would not be possible to fly a modern twin-engined aircraft and operate radio equipment on such a flight without the help of a skilled navigator. Amelia Earhart and her navigator, Fred Noonan, perished when they failed to locate the tiny dot of an island, 2 miles long and ½ mile wide, in the middle of the Pacific. During the last leg of their journey, a 2556-mile flight from Lae in New Guinea, they ran out of fuel whilst flying up and down along a line within the area where their target lay.

Harold Gatty, who had accompanied Wiley Post as his navigator in July 1933 on a global flight at northern latitudes (some 10,000 miles less than the equatorial distance Amy intended to make), was present whilst she was with Weems on the course, and one speculates on the possibility that she might have tempted the Tasmanian into becoming her navigator on her proposed flight. We shall never know, since her plans for the flight were scuppered when she next met Dupré.

She returned home in June and had a blazing row with the hotelier when she visited Paris. She found that he had decided to make a prolonged continental trip in his new Caudron aeroplane, which meant that it was not available for her to use. It appears that time had run out for Amy, and that her would-be suitor had begun to cast his marriage net in another direction. Amy's father urged her to not to be too hasty in falling out with the man who was in effect keeping her in a lifestyle that she could no longer afford. A succesful global flight might well have restored her fortunes as an aviatrix and made her financially independent, but without his backing there was little chance.

Dupré made one last throw of the dice in order to secure the prize he coveted most when he invited Amy to stay with him at his stud-farm for a weekend in July. She agreed to go but, as she subsequently told her father, it was with the utmost difficulty that she 'avoided the purpose for which she had quite obviously been asked'. It seems that during her stay, Dupré would have settled for her being his mistress, but she would have none of it.

Around this time she was romantically involved with one of Dupré's colleagues, and went to Cannes with him whilst Dupré himself was staying at the Carlton. Little suspecting that Dupré already knew of her clandestine holiday, she visited him at the Carlton alone whilst she was there. Much to her surprise, he confronted her about what he considered to be her disloyalty, and told her that by her action she had effectively stepped out of his orbit. There were no raised voices and everything was conducted in a smooth and civilised manner. However, he was quite firm and told her that they were no longer partners.

In December 1937 the French hotelier was seen boarding a liner bound for Canada with an attractive young Hungarian girl on his arm. A little later the marriage of M. François Dupré and Mlle Anna Nagy was announced in the press.

Chapter Eighteen

Losing Altitude

It has been said that the most poignant legacy of abandonment is compulsive wandering from person to person and from place to place. Although Amy Johnson was never abandoned as such by her family and friends, the saying does reflect with a great deal of accuracy the peripatetic existence she resumed after she lost the patronage of François Dupré. Now that record-breaking flights had lost their public appeal she became acutely aware that she needed to find an independent source of income if she were to maintain the lavish lifestyle that went with her celebrity status. Various men came into her life but she was never sure whether it was her they really wanted or merely the cachet of having an association with her fame. In one of her letters to her parents she made no secret of the fact that she would 'keep a look out for a nice rich husband', but sadly she never found one.

During the early autumn of 1937 Amy entered the Champneys Health Clinic at Wigginton near Tring, and whilst she was there taking a 'rest cure' she made two important decisions. She decided to rent out her Gate House residence and leave London permanently, so that she could live more cheaply in the country. Coincidentally, two of her best friends, Dr W. A. 'Sandy' Hislop and his wife Katherine (née Mollison – one of Jim's five aunts), offered her the use of their small cottage at Haddenham in Buckinghamshire for a few weeks. Her brief stay in its idyllic surroundings convinced her that she was doing the right thing in moving away from the capital. At the same time she decided that she needed a new interest in her life, and whilst she was at the clinic she visited the London Gliding Club at

Amy looks rather apprehensive whilst being briefed for her first hilltop launch in the nacelled Dagling glider. (Edward Hull/London Gliding Club)

Dunstable. During that visit she became fascinated with the sport and took up gliding seriously as one of her hobbies. It was a case of starting from scratch with ground slides in the primitive Dagling Grasshopper trainer, until she progressed to bungee-assisted launches from a hilltop in the real thing – sailplanes. Like most pilots who move from powered flight to soaring silently over the landscape – to fly like the eagle – she experienced the exhilaration of a new freedom.

Amy came to love this area of the Chilterns, and eventually rented a furnished cottage called Monks Staithe, a former vicarage situated opposite the local church and churchyard of St Mary's in Princes Risborough. It was at this half-timbered residence, which she shared with her dachshund Tina and housekeeper Jessamine James, that she was able to entertain friends and enjoy a relaxed lifestyle. Fortunately, through her friendship with some of the staff at the riding stables nearby, she was able to indulge in her love of horse-riding quite cheaply, since she was allowed to exercise their horses on their morning canter. Local people remembered her riding with the hounds of the Old Berkeley Hunt and pictures taken at the time show her elegantly dressed for the event in black bowler hat and riding jacket mounted on a bay stallion.

Living at Chestnut Farm House in Monks Risborough, not far from Amy, at that time was the Fletcher family, one of whose members by a strange twist of fate was to play a critical role in the last stages of her life. Walter Edmund Fletcher, son of an Anglican clergyman, was a young naval officer who had recently been a member of the 1934 polar expedition, which set out to conquer the North West Passage, and which made several important discoveries. As a consequence of that mission he had been made a Fellow of the Royal Geographical Society, much as Amy had been before him for her Australian flight, and it is most probably through this shared interest that they first became acquainted. Their two lives were to become inextricably linked.

In spite of living away from the capital, Amy was a constant visitor to it during this time, mainly through financial need. She was still regarded as newsworthy by the media and as a result she made an appearance on

Amy rode with the Old Berkeley Hunt whilst she was living at Monks Staithe during 1937/38. (The Amy Johnson Estate)

television, something rare in those days for it was still very much in its infancy in the early part of 1938.* She was also the aviation editor of the *Daily Mail*, producing eye-catching articles such as 'Why not a women's air force too?', as well as writing for *Thomson's Weekly* and various Sunday national newspapers such as the *Express*, *Graphic* and *Dispatch*. She was also busy writing her own book, *Skyroads of the World*, which was published in 1939, as well as a memoir of her early life, *Myself When Young*. Amy was not a particularly gifted writer but she enjoyed the peace and tranquillity involved, so different from the frenetic lifestyle she had been living. In an article for *Lilliput* in 1939 she looked back on those halcyon days at Monks Staithe by recalling her time there with:

> I am left undisturbed – 'She's busy writing,' explains the dear old woman who 'does for me', in that sort of awed whisper which is supposed to convey a great understanding for an author's birth-pains. Most proud I am to see an article or a story of mine in print. Much more proud than I ever was to see some sensational story I had never written splashed across a newspaper's front page. That somehow never seemed really me.[30]

When Amy was behind the wheel of a car she loved to drive with her foot well down on the accelerator, and it is not surprising that her name was associated with the 1938 Monte Carlo Rally. She entered with the intention of driving a French Talbot-Darracq saloon in the 3-litre class, but for some long-forgotten reason never competed. As it transpired, she settled for the more modest women's Paris–St Raphael rally. It was her first big motoring event and she found that the 650-mile drive, much of it on icy roads, demanded skills that she had yet to acquire. There was some recompense for her however in the Concours d' Elegance at Cannes, where she gained the attention of the French paparazzi as they scrambled to get pictures of *la celebrée aviatrice anglaise* looking her most attractive in a chic spring outfit.

Her first taste of rally-driving whetted her appetite for more and she pursued the sport with her usual enthusiasm, but without ever achieving the acclaim she received as a flyer. She entered for some of the more important UK events, such as the May meeting of the Shelsley Walsh Hill Climb, where she achieved the third fastest time of the day driving a Frazer-

---

* The fee for her BBC TV appearance was what might seem to be a lowly £10, but at today's values it would be more like £1000.

Nash BMW against all comers. In the following month she took part in the five-day Scottish Motor Rally, but she achieved only moderate success.

Whilst her competitive driving aroused little real public interest her private motoring did raise a few eyebrows, for on two occasions, in High Wycombe and Worksop, she was fined for speeding. She loved fast cars, and her choice of a Ford V8 roadster was heavily influenced by the success that had accompanied the marque in the Monte Carlo rallies of the thirties. With its powerful eight-cylinder engine and its distinctive acorn headlamps, it had achieved a certain degree of notoriety in the States when it was used as a get-away car by Bonnie Parker and Clyde Barrow, the renowned gangster couple who have now passed into cinema history. Just six weeks before the pair were gunned down in a black sedan, Clyde had written to Henry Ford with a cheeky note eulogising the company's latest product. His roughly scrawled, semi-literate note read:

> While I still have got breath in my lungs I will tell you what a dandy car you make. I have drove Fords exclusively when I could get away with one. For sustained speed and freedom from trouble the Ford has got ever other cars skinned, and even if my business hasn't been strickly legal it don't hurt anything to tell you what a fine car you got in the V8.[31]

Although the decree nisi for Amy's divorce came through in the early February of 1938, she never quite managed to throw off the trauma of her failed marriage. In a letter to her parents a few months earlier, she had told them, 'Jim is being sued for £4000 income tax and apparently intends to try to get some from me. What a hope. If he sues me I shall just put in a large claim for maintenance.' A few weeks later she wrote, 'I have to appear at Jim's income tax case next Wednesday. Crocker's have let me in for that and I am very angry with them as it was quite unnecessary and is very embarrassing for me.' Amy knew that Jim Mollison was unscrupulous in money matters and notorious for leaving a trail of debts behind him, and no doubt she was more worried by his threat to involve her than she cared to admit.

The ever faithful Peter Reiss came to the rescue. He was still carrying a smouldering torch for Amy and he recalled how desperately hard up she was in 1938. It came to the point where she was too embarrassed to visit London pawnbrokers herself in order to sell some of the jewellery she had been given after her Australian flight, for fear of being recognised. As a consequence, she asked him if he would do it for her. PR dutifully did the rounds touting for a quote on various pieces of jewellery, until he finally settled for an establishment in Victoria, which seemed slightly more

Mollie, Amy and her solicitor outside the Law Courts at the time of her divorce from Jim Mollison in August 1938.
(Hull Daily Mail)

generous than the rest. When he reported back to Amy she was shocked at the cut-throat prices she was being offered. Unfortunately, she had no alternative but to accept.

The ignominy of the sale was further compounded for Amy when her friend returned to the shop with the goods only to be told that since the jewellery belonged to a woman, then the lady concerned would have to be present to sign 'the usual forms' before they would pay out. Early the next morning, PR and a heavily muffled Amy, with her hat pulled well down on her forehead, stood before the assistant, who presented the form and asked her to sign her full name. It was obvious to PR that the man behind the counter had recognised her immediately. Amy signed 'A. Johnson', thinking that that would suffice. 'We do need your full name, Madam,' he said, and then tactfully added, 'Perhaps I can help by suggesting Amelia, would that do?'

'That's fine,' said Amy. She collected the cash and they quickly left the shop.

When German troops marched across the Austrian frontier on 11 March 1938 in an annexation of that country which Hitler euphemistically termed *der Anschluss*, there was little doubt, except amongst a few willfully blind politicians, that a major war in Europe was now inevitable. One of the consequences of the German invasion was the announcement by the British government of its far-sighted Civil Air Guard scheme, which was due to start up on 1 September. It was part of a national defence programme in which men and women between the ages of eighteen and fifty would be trained as pilots as a supplement to the RAF, at some sixty local flying clubs

Amy arriving at the Hooton Park aerodrome on 9 September 1938 in Tipsy B,
G-AFCM. *These small two-seater monoplanes were used by the Civil Air Guard for
giving instruction to pupils.* (E.J. Riding)

throughout the country. Effectively, it was the equivalent of the Territorial
Army and Royal Naval Reserve. Owing to a quite unexpected rush of
applicants – some 35,000 in the first few months – an appeal was made for
qualified instructors. This did not escape Amy's attention.

As soon as she heard of the proposed formation of the CAG she sought
the help of her old friend Francis Shelmerdine, Director of Civil Aviation,
looking for an opportunity of playing a major role in it. Although she
managed to get an interview with the Air Minister, Sir Kingsley Wood, who
was outwardly polite and sympathetic, all he would offer her was the post of
junior operations officer at £5 a week. It was an insult, and whilst she
managed to keep cool during the interview, she seethed in anger at his
derisive response. She was hurt professionally because she felt that
officialdom regarded her as nothing more than a 'circus flyer'. However, the
rebuff did little to deflect her from her aim of making a serious contribution
in aviation, all the more so now that war was looming.

In spite of the seriousness of the political situation in a country on the
brink of war, with volunteers digging trenches in Hyde Park, and public
buildings being protected with sandbags, life continued much the same for
Amy. She consoled herself by occupying her time gliding, a pastime she
came to love. Reminiscing on that period she recalled:

> I have memories which no one can take away and I have learned just one
> or two simple truths which give me happiness. I have a few friends I
> dearly love, a few books, a tiny cottage and garden which give me a calm
> content. I have a sailplane up at Long Mynd which I fly every weekend
> among the clouds and blue sky in a silence broken only by the sighing of

the wind in the wings, the laughter of my friends, and the occasional cry of a bird.[32]

The gliding fraternity soon discovered, much as the hangar staff at Stag Lane and others had done during the early days of her powered flying, that Amy was not one to shirk the menial tasks. She did not stand on her dignity or hide behind her femininity, but was prepared to be 'one of the boys'. Launching in those days usually required four strong men pulling their hardest on the ends of the two bungee cords which formed a 'V' with the hook on the nose of the glider, whilst another two lay prone behind the craft, holding it back until the signal to release was given. Pictures taken at the time show Amy flat on her stomach clutching the feet of the member in front of her, who was holding onto the tail of the machine until it was released. It was this side of her character that won her many friends.

Amy helping to steady a Wren sailplane which is about to be bungee-launched from Dunstable Downs in 1938. (Edward Hull/London Gliding Club)

One of the founder members of the gliding club at Scarborough was Fred Slingsby, a Yorkshireman who went on to establish his own sailplane factory at Kirbymoorside in 1931, the first of its kind in the UK. As soon as he knew of Amy's involvement with the sport he offered her the permanent loan of one of his single-seat Kirby Kite sailplanes. From then on she was to be seen arriving in her Ford roadster at clubs all around the country. She would arrive at Dunstable, Long Mynd, Sutton Bank or Camp Hill in Derbyshire with her dismantled sailplane tucked neatly into a trailer. The Slingsby product was a fully aerobatic plane, and Amy loved nothing better than to be catapulted from a hilltop into an up-current and put her machine's gracefully tapered wings into a loop over the patchwork-quilted landscape beneath her. She was once quoted as saying: 'Sailing smoothly and gently through the sky in a glider was the next best thing to graduating as an angel.'

At the beginning of August the lease ran out on Monks Staithe, and reluctantly she was forced to move yet again. For some obscure reason she took up temporary residence in the Rossley Manor Country Club at Andoversford, near Cheltenham. Her involvement in competitive motoring enabled her to add yet one more string to her bow, for she was invited to be editor of the magazine *Lady Driver*. This meant regular trips to London to discuss her work as a writer for the newspapers and also for what she termed 'voice production' lessons.

In an age when BBC received pronunciation was the norm for the circles in which Amy moved and when girls were referred to as 'gels', regional accents were deemed to be rather *infra dig*. Amy had never quite managed to eradicate the slight touch of Yorkshire in her voice. She was very conscious of it, especially on one occasion when she happened to be visiting one of London's best-known restaurants with an American friend. The rather risqué and waspish cabaret artiste Florence Desmond was performing that evening, and her usual act consisted of impressions of stars of stage and screen. She would delight her audience with impersonations of Marlene Dietrich, Katherine Hepburn and Garbo, each one given to perfection. Then came a spot when the audience would usually challenge her by shouting out a request and give a name for her to caricature. Seeing Amy and her friend seated at a table nearby, someone shouted, 'Amy Johnson! We want Amy Johnson!' Amy's heart missed a beat. Miss Desmond was not exactly a fan of Amy's, owing to an altercation between her late husband, the aviator Tom Campbell Black, and Jim Mollison during the run-up to the MacRobertson air race in 1934. She gave a devastating impression of Amy with an exaggerated Yorkshire accent, which caused the victim to squirm and turn to her friend and ask, 'Does my voice sound like that to you?'

'Not in the least. Let's get out of here,' came the swift reply. Out on the street she thought, 'He's only saying that because it's easiest. No man tells the truth when he's with a woman. It takes a woman wilfully to hurt another woman's feelings.'[33]

Amy suffered a further public embarrassment when the newspapers splashed pictures of Jim Mollison and his new bride across their front pages.

The summer of 1938 had seen Amy's ex-husband living the bachelor playboy existence as irresponsibly as ever. He worked his way through France much as he had done ten years earlier when he was discharged from the RAF, only this time he was an internationally known figure. With the proceeds from the sale of the Bellanca aeroplane, which he sold to the Franco government for use in the Spanish Civil War, he was to be found in the casinos of Le Touquet spending as if there were no tomorrow and hitting the bottle as hard as ever. Part of the time he spent on the Riviera and whilst he was there he met the thirty-four-year-old heiress and socialite, Phyllis Louis Verley Hussey. Their first meeting took place in August of that year, probably at the Carlton in Cannes, for apart from being one of Jim and Amy's old stamping grounds it was, according to Phyllis's daughter, her mother's favourite summer idyll.

The willowy blonde divorcée, a member of the distinguished Verley family, was reputed to have an income from her father's legacy of £20,000 a year; with that kind of wealth and her good looks, she was every unattached male's dream-ticket. She was a prominent figure in Jamaican high society. Her first marriage, in the early twenties, had been to Lt Cdr Thomas Hussey, RN, a man who was to become equerry to Lord Mountbatten during the Second World War. Fleet Street was hammering out articles with captions reading, 'Jim Mollison to wed millionairess after whirlwind wooing', whilst at the same time reporting that he was planning a record-breaking round-the-world flight. One cannot help suspecting that he saw Phyllis's money as a ready source for such an undertaking.

Just how serious Mollison's claims were about such a venture are open to question. Admittedly, there had been other contenders, such as Amy and Beryl Markham, who had both declared an interest since Amelia Earhart and Fred Noonan's death in the Pacific a year earlier, but in each case they had failed to secure the financial backing necessary. The eccentric American millionaire, Howard Hughes, with the help of four crew members and the most up-to-date navigational equipment installed in his twin-engined Lockheed passenger aircraft, had circumnavigated the globe in July of that year and halved Wiley Post's record time. But again it was made north of the Tropic of Cancer and therefore some 10,000 miles short of a true equatorial distance. The prize was still there waiting to be claimed.

The Forum Club for professional women had been established in Grosvenor Place, London, soon after the First World War, at a time when women were just beginning to gain their political and social emancipation. In many ways it resembled its male counterpart in having the same quiet atmosphere, with a marbled hall, plush carpets, library and liveried flunkeys, but unlike men's clubs it was less rigid in its rules and magnanimously welcomed those of the opposite sex. It was a place where at any time one might bump into someone as important as the Prime Minister's wife or a celebrity from stage or screen; where one could play billiards or bridge, attend lectures, drink cocktails and chat with a friend at the bar or where women members could relax in the beauty parlour under the hairdryer. Membership was drawn from all sectors of society and it catered for a variety of activities and interests, with the result that within its structure there were twenty-four different societies. One of these was the Aviation Society.

Amy had been a visitor to the Forum Club since her first introduction to it in 1931, and it became a venue where she met up regularly not only with other women flyers, but also with women who had political influence. At the Aviation Society's annual dinner in December, a motion was passed for the society to become the Women's Aero Club, rather than just being a society within the Forum Club. A committee was formed, with Princess Marie Louise as president (she was the first member of the royal family to fly) and Commandant Mary Allen, Amy Johnson, Pauline Gower and Dorothy Spicer as committee members. Their immediate task was to increase their membership and to examine the role that women might play in the event of a national emergency. As a panel of experts, their influence was to have a far-reaching effect upon the practice of women ferrying aircraft from the factories to the operational squadrons of the RAF when the war finally did break out.

Amy continued with her interest in competitive motoring and teamed up with Dorothy McEvoy in January 1939 to drive a Ford V8 22-horsepower saloon to enter for the Ladies Cup in the Monte Carlo Rally. They chose to start at John o' Groats (one of the eight permissible departure points) for the gruelling nightmare of a course. It meant three days and four nights of continuous hard driving through the icy winter conditions of northern climes until they reached the sunshine of the Riviera. In spite of the fact that Amy's co-driver was one of Ford Motors' most skilful lady drivers, the 5000-franc prize eluded them, but they put up a creditable performance. Amy always maintained that rallying was more exciting than flying although, thankfully, not as terrifying as flying over vast stretches of ocean.

The fast driving to which Amy became accustomed whilst rallying on public roads might well have been a factor in the road accident in which she was involved soon after her return from Monte Carlo. She was driving

Dorothy McEvoy and Amy Johnson with their Ford V8-22hp Saloon which they entered for the 1939 Monte Carlo Rally. (The Amy Johnson Estate)

home from a meeting of the Forum Club early in February (she was now living in a rented cottage called Old Rowley in Stoke Orchard, near Cheltenham) when she collided with another car near Witney in Oxfordshire. She was fortunate in being given no more than a caution for driving without due care and attention and, whilst she was not seriously injured, she damaged her right knee badly enough to immobilise herself for several weeks.

There is no doubt that the shock of the accident frayed Amy's nerves and she decided on a sea voyage to recuperate, sailing for New York in the luxury liner *Ile de France*. Pictures taken as she was about to board the ship show her, looking rather wan and forlorn, being pushed along in a wheelchair. Her finances as well as her spirit must have been pretty low around this time, for in a letter to her father she told him that she might have to ask him for a small loan, and that she proposed to sell a pair of earrings that François Dupré had given her. It seems that throughout the whole of her life the only man whom Amy could ever entirely rely upon was, in fact, her father. The rich knight in shining armour never rode into view.

When Amy sailed back from the USA at the end of the first week in April, it was to a country teetering on the edge of war. Neville Chamberlain had returned from his meeting with Adolf Hitler at the Munich Conference the previous September, when it was obvious that war was inevitable. As the Prime Minister stepped down from the plane at Heston holding a worthless document, he assured the nation, in a wavering voice bereft of any confidence, that it would be 'peace in our time'. Few believed him, for they knew that it was now a matter of staving off the enemy for as long as possible whilst the country rapidly rearmed.

RAF pilots were being trained in large numbers at the elementary flying training schools dotted around the country, where the ubiquitous, yellow-painted Tiger Moths were to be seen processing the airfields as their instructors took their pupils through their 'bumps and circuits' routine. It was a fairly risky occupation for the instructors, and it was not unknown for them to be killed or seriously injured whilst teaching others to fly.

One such was Amy's old flying instructor, the man who had told her that she would never make the grade as a pilot when she began her flying lessons at Stag Lane in 1928. Froude Ridler Matthews had rejoined the RAF with the rank of Flying Officer, as war became imminent. Whilst he was instructing at the Fleet Air Arm Training School at Gravesend, in May 1939, teaching RN midshipmen to fly, he was killed. Several of the Tiger Moth trainers were circling the airfield on that day when two collided whilst making a landing approach. Eye-witnesses working in the fields nearby reported seeing and hearing the two planes touch, whereupon they nose-dived and fell into a small copse. One of the instructors was killed outright and his pupil in the rear cockpit badly injured. Matthews was pulled from the wreckage of the other machine seriously injured and unconscious, but as he was being put into the ambulance he came to and said, 'It wasn't my fault.' With almost his last breath he asked how his pupil had fared and how the two men in the other plane were. He died later in Gravesend hospital.

The armed services were not alone in preparing for hostilities; small charter airlines were recruiting extra pilots to fulfil lucrative contracts with the military to conduct army co-operation flights. It usually meant a civilian pilot flying an aircraft on a set course so that anti-aircraft gun batteries could test their accuracy in finding the range and height of an enemy machine. Night flying, for which these pilots were paid a bonus, was also carried out for the benefit of the army searchlight crews to practise beaming in on enemy raiders.

One airline which became engaged in this work was Portsmouth, Southsea & IOW Aviation Ltd, a small company with a dozen or so aircraft which was known to locals as the 'Pip, Squeak & Wilfred' oufit, after the well-known cartoon characters in the national press at that time. The company was recruiting extra pilots for the military work and Amy seized the opportunity to get back into flying. She joined the airline in June. She was not the only woman pilot to be employed by them, for at the same time they recruited Joy Davison, a young woman who had been giving pleasure flights using her own DH Moth from a field on Hayling Island. Both women were eventually to join the Air Transport Auxiliary and ultimately to share the same fate.

For the next three months Amy was quite satisfied to be earning £1 a day, plus another 10 shillings per hour for daylight flying, although it irked her that the night flying, which attracted an extra £1 per hour, was restricted to the male pilots. Apart from the army co-operation work, she was also busy flying holidaymakers in one of the company's four silver, blue and white liveried Airspeed Couriers from Portsmouth to Ryde in the Isle of Wight. These were flown right up until the outbreak of war, on what was popularly known as the Spithead Express. Not long after Amy started, the *Daily Mail* ran an article which read, 'Folks, you've got the chance of being flown by a world-famous pilot for five bob a time.' Passengers were delighted to be flown by someone who was still considered to be a national heroine, and no doubt if she had charged for the many times she was asked for her autograph, she would have added substantially to her income.

The company also ran a combined air and coach tour of the Isle of Wight, leaving Portsmouth airport at 10.30 a.m. It landed at Ryde to connect its passengers with the 11 a.m. grand tour of the island, from which they returned in time to catch the 6.40 p.m. Courier flight back to the mainland – all for the princely sum of 14s 6d! Some people who took those pre-war trips remembered a well-known national firm of chemists advertising their package for air travellers, which contained chewing gum, smelling salts and a sick bag. They recalled that engines were kept running as passengers hurriedly exited from the aircraft to allow another batch to board. Seatbelts were conspicuous by their absence and, on the rare occasions when there were only a few passengers, they would be directed by the pilot to sit exactly where he or she wanted in order not to upset the trim of the aircraft in flight. In spite of the rudimentary comforts of the airline during the seven years in which it operated, it could boast of an impressive safety record, carrying 220,000 passengers without death or serious injury.[34]

Amy moved from her lodgings near the airport not long after she joined the airline to rent what she described as 'a lovely bungalow' at Old Bosham, near Chichester, and whilst she was there she became involved in a craze for sailing. As with every other sport she took up, there was never anything but total commitment. She was soon writing home to her parents, telling them of the thrill she got from her new sport; that it was 'much more exciting than flying' and that she was contemplating a sailing holiday on the Clyde with two of her gliding friends.

Gerald Edwards, a senior pre-war instructor at the Midland Gliding Club and fellow gliding enthusiast Cecil Reilly, recalled how they shared costs with Amy on that yachting holiday, starting from Helensburgh on the Clyde, just before Hitler spoilt everything. Both men were some ten years younger than Amy and there was never any question of a romantic involvement, but both were deeply impressed by her level-headedness and the way in which she shared equally in the tasks throughout the fortnight. They also recalled a time when the weather closed in on them whilst they were with her at Long Mynd during a gliding weekend. Amy got fed up waiting around and said, 'Come on, I've got a good idea. We'll go to Blackpool and ride the big dipper.' They immediately piled into her open-top Ford and sped off to her sister's home at the resort where they spent the night. Mollie's husband, Trevor, who at that time was the Town Clerk, wanted to take all three of them and his wife out to a well-known local restaurant for the evening, but Amy declined, shunning the publicity that that might involve. However, they remembered her sharing with them the delights of the Pleasure Beach that evening.[35]

On 31 August 1939, three days before the outbreak of hostilities with Germany, an air navigation restriction order came into force, which meant that all civilian aircraft, apart from some operating with services to the Scottish islands and the Scilly Isles, were grounded and placed at the disposal of the Air Ministry. The aircraft and their civilian pilots from eighteen air companies now came under the control of an organisation known as National Air Communications, with its major base at Cardiff. The commandeered aircraft were speedily repainted with drab camouflage colours and used in a variety of communication flights in which VIPs were flown within the UK and to the continent. Other tasks included carrying freight and running a blood delivery service to the British Expeditionary Force in France.

Amy was directed to report to the airfield at Heston and await further orders whilst the rest of the personnel working for P S & IOW Aviation were moved on 1 September to Cardiff's municipal aerodrome at Pengam Moors, just 2 miles to the east of the city centre. Such was the air of secrecy that surrounded their move on the morning that they left Portsmouth, that their destination was only revealed when the long, straggling column of vehicles halted at Birdlip Hill. They arrived in the evening and the next day was spent unpacking the vast quantities of equipment that had accompanied them, and digging their own air-raid shelters. The men were progressing with their excavations on the Sunday morning, whilst Joy Davison and the manager's wife were rustling up some food *al fresco*, when a small scruffy boy from the nearby council estate ran around the streets screaming at the top of his voice, ' 'Tis War, 'tis war!'

Amy soon discovered that the Heston posting was nothing more than a bureaucratic mix-up, and within a few days she was redirected to join the rest of her colleagues in Cardiff, whereupon she went into shared accommodation with two of the other NAC pilots. In a letter written soon after her arrival in Wales, she said, 'This waiting about is beginning to get me down. The trouble is we have to be at the airport the whole time in case we're wanted.' The army co-operation work continued, together with some anti-submarine patrols out into the Bristol Channel, but the work was sporadic and she soon became bored and frustrated.

Although she played squash to release some of her tension, she missed the activities which she had grown to love – the freedom of gliding, horse-riding, sailing and competitive motoring. All these were now denied to her by the restrictions of a country at war. Those who knew her well also felt that she missed being married, if not to Jim Mollison then at least to someone, and the enforced inactivity only gave her time to brood on this aspect of her life. She was desperately unhappy. She had achieved so much in her career and now the war seemed to exacerbate the frustration she felt. To make matters worse, she was miserable in Cardiff - 'a detestable place' is how she described it – but this was no doubt due in part to a fracas she had with the local police soon after arriving there.

After war was declared blackout restrictions were imposed not only upon all public and private buildings but also upon motorists, with the intention of denying enemy aircraft knowledge of their position during a night raid. This meant that drivers were required to drive on sidelights only if they did not have proper shields over their headlights. Amy happened to be following close behind a car which was using full headlights on the road from Cardiff to Cowbridge, when they were waved down by a local policeman at the village of St Nicholas. Whilst the first driver was being spoken to, Amy was alleged to have shouted, 'How much longer are you going to keep me here?' This annoyed the over-zealous constable, who then turned his attention to the one questioning his authority. When he demanded to see Amy's driving licence, she is purported to have given him a hard time and replied, 'I'm damned if I will, you fool,' and began to drive away, whereupon the irate bobby jumped onto the car's running board and attempted to snatch the ignition key. Amy eventually stopped her car and she and her passenger were told to get out. Both refused and she began to drive on, again with her hostile hitch-hiker standing on running board, only this time he threatened to smash the windscreen with his truncheon if she did not stop.

In retrospect, the whole incident takes on the character of an amusing scene from an early Ealing comedy, with the village copper being driven at speed down the road whilst hanging on for dear life. Apart from her basic unhappiness at this time, the incident highlights her pent-up anger at what she might consider to be an example of overbearing male authority, something from which she suffered throughout her flying career. Maybe when she looked into that policeman's face she saw a kaleidoscope of Hans Arregger, Jim Mollison, Herbert Travers, and all the other men against whom she had had to battle. Nevertheless, she was ordered to appear in court and in the witness-box she apologised for using the word 'damn', although she denied using another expletive which the constable described as 'unbecoming of a lady holding her position'. She was fined a total of £6.7.6d on four counts and her licence was endorsed.[36]

Whilst Amy was with NAC, she was called upon to fly military officials and VIPs to the continent. As we shall see, these flights were to give substance to many confusing and unfounded rumours which were to circulate a year later. On one such trip she was detailed to fly a senior RAF officer from Shoreham in an unarmed Miles Falcon monoplane to an emergency airfield at Rheims, where some of the Hawker Hurricane and Fairey Battle squadrons were based. A journalist covering that part of the 'phoney war' period happened to be present in the bar of the Hotel Lion d'Or when a buzz of excitement went around the room as soon as Amy entered. 'I say chaps, look who's here. I do believe it's Amy Johnson.' Before long she was surrounded by a group of admiring young RAF pilots, plying her for autographs and raising their glasses to toast her good health with champagne. It was to become a familiar story wherever she went during her visits to wartime airfields.

Whilst unarmed flights in civilian aircraft to northern France were reasonably safe from enemy interception during the early months of the war, they were by no means risk-free, and it is interesting to note that on 5 January 1940 Amy amended her will. The date is quite significant and one wonders if she had some kind of premonition when she wrote it. Within a few days it was announced that she was to be posted to the aerodrome at Woodley, near Reading, to continue with her army co-operation flying. Soon after she left Cardiff she wrote, 'I am happier now than I have been since the war started. I am my own boss here and am in charge, which is so much nicer. Also I get every weekend off . . . I like the people much more and have a lot of friends.' However, in less than a month the news came through that the flying she was employed on was to be taken over by RAF personnel, which meant that she was virtually out of a job. Amy viewed the whole scenario as just one more example of male prejudice against women, since most of the men with the company were assimilated into the RAF to continue with the work. She was now forced to consider the few options that were left open to her.

The Air Transport Auxiliary came into being as the result of action taken in 1938 by the Director General of Civil Aviation, Sir Francis Shelmerdine. His plan was to utilise those airline and civilian pilots who, for reasons of age or physical fitness, would be unacceptable to the RAF, for service on national communications work. By the end of June 1939 he had managed to get Treasury approval for the formation of a Civil Reserve of Pilots, one which was initially under the control of Imperial Airways and British Airways. (These two organisations amalgamated on 1 April 1940 to become British Overseas Airways Corporation.) Finally, it was decided to appoint Gerard d' Erlanger, a director of British Airways, to recruit suitable candidates from those pilots holding 'A' Licences. So it was that the plan was approved and on 3 September 1939 the ATA was born.

It was proposed that civilian pilots between the ages of twenty-eight and fifty who had a minimum of 250 hours' flying experience should be considered, and that they should report for final selection at Whitchurch, near Bristol. If suitable they would be paid at the same rate of pay as junior officers in BA – approximately £350–400 per annum – and be issued with a similar uniform. Their duties would include the transport of dispatches, mail, medical supplies and important civilian personnel. In addition, they would co-operate with the police and fire brigades.

The ferrying of aircraft from the manufacturers' airfields to the operational squadrons was not, initially, part of the ATA's remit; since the end of the First World War this duty had been carried out on an *ad hoc* basis by RAF squadron pilots themselves. However, during the pre-war expansion of the RAF a scheme was inaugurated in which thirty-two RAF pilots, operating from two ferry pools at Hucknall and Filton, shared the ferrying of aircraft from the factories to the Elementary Flying Training Schools, and into and out of several storage units. This continued throughout 1939 until the pressure on these service pilots became such that the work was partially handed over to the ATA, in what were termed mixed ferry pools. Before long the whole of the work was given over to the ATA and it eventually became their main role.

Inevitably, with the creation of an all-male ATA, the voice of a small but significant band of women pilots was soon demanding to be heard. Apart from Amy, there were many other lesser-known but equally capable women such as Pauline Gower, Dorothy Spicer, Mona Friedlander and Mildred Bruce, who had already carved a career for themselves in aviation. They were determined to contribute by serving as pilots, and, moreover, on equal terms with the men. Women pioneers such as Amy Johnson, Jean Batten and Amelia Earhart had already demonstrated that they could compete and hold their own in what had once been considered a male domain.

The most influential person in securing a role for women in the ATA during the latter part of 1939 proved to be Pauline Gower. Not only was she a very competent pilot, having gained her 'A', 'B' and navigational licences and with over 2000 flying hours to her credit, but she was also the daughter of the solicitor and MP, Sir Robert Gower. There is no doubt that she had the advantage of some of her father's political clout behind her when she approached Shelmerdine soon after the outbreak of war with a proposal for the formation of a women's section of the ATA. After a great deal of bureaucratic huffing and puffing, she was given approval to recruit eight women pilots to ferry open-cockpit Tiger Moths from the de Havilland factory to stored reserve in Scotland. It was a task that the RAF was only too glad to shed. And so, on 1 January 1940, the first entirely civilian ferry pool was established at Hatfield.

Many people must have wondered why Amy, who was at the time still engaged on army co-operation duties with P S & IOW Aviation at Cardiff, was not chosen to form the women's section of the ATA. After all she was an internationally known figure and on very good terms with Shelmerdine, (he was best man at her wedding in 1932). The answer clearly lies in a minute from Shelmerdine to the Director of Civil Aviation Finance dated 23 September 1939. After outlining the background to the proposed women's section, he went on to recommend Pauline Gower for the job of forming it, and then added, 'She has never been a *stunt pilot* with all the publicity that is attached to that role.' One can only speculate on whether the Cardiff incident and its subsequent coverage by the press influenced Shelmerdine's decision.

Chapter Nineteen

ANYWHERE TO ANYWHERE

I n March 1940 Amy Johnson's contract with Portsmouth, Southsea & IOW Aviation was terminated and she was left jobless. She became totally frustrated to find that the fame which had been heaped on her over the past ten years, was now actually working against her in her desire to make a worthwhile contribution towards the war effort. The appointment as the head of the women's section of the ATA had virtually slipped through her fingers, and whilst Pauline Gower implored her to join with the nine women who already made up the ferry pool at Hatfield, Amy felt that the work they were initially being asked to do was derisory.

During this period Amy began to write prolifically, much as she had done when looking for sponsorship for her Australian flight, to people of influence offering her services to her country in whatever way possible. She had a considerable number of political contacts through her membership of the Forum Club, one of whom was the Prime Minister's niece, Valerie Cole, who happened to reside at 10 Downing Street. Amy became friendly with her, and it is most probably through her having the ear of the Prime Minister that Amy gained an interview with Lord Vansittart somewhere around this time.

The tall, urbane diplomat, Robert Gilbert Vansittart, had enjoyed a successful career before the outbreak of the Second World War, having served as Permanent Under-Secretary of State for Foreign Affairs for eight years right up until 1938. His biographer described him as 'lean in appearance, with a darkish complexion and high oriental-like cheekbones' and said that in his demeanour 'it was difficult to resist the pervasive evidence of his charm'.[37] Unfortunately, he tended to be outspoken to the ministers he served, and as a result had fallen foul of his Foreign Secretary, Anthony Eden, who eventually sacked him. He disappeared into a backwater until he was appointed as the Chief Diplomatic Adviser to the Foreign Secretary from 1938 to 1941,* a period which coincided with the formation of an organisation designed to train and equip agents to sabotage the enemy in occupied territories in the event of the fall of France.

No records survive of the meeting between Amy and Vansittart, but it is known that she offered her services to the nascent Special Operations Executive, an organisation in which he had a great deal of influence. She had a good working knowledge of French, knew the country well and was competent enough to fly agents in and out of it. According to Amy's biographer, Constance Babington Smith, 'She begged him to take her on for some dangerous mission in the Secret Service.'

One wonders just how this interview went and exactly how Vansittart perceived Amy Johnson. Here was a man terrified of flying who had never even learned to drive a car, listening to a woman who revelled in the extremes of both those activities, and who was now willing to throw her life away for her country if necessary. He must have admired her courage, even if he did spurn her offer as being made out of naivety.

In what must surely have been a cry of despair, Amy let it be known that she was even willing to drive a lorry as part of her contribution to the war effort if that is what it took. This was not as unimaginable as one might suppose, since another famous woman pilot, Jean Batten, faced with a similar problem, settled initially for being a wartime driver with the Anglo-French Ambulance Corps, and when that folded after the fall of France she finished up working as an inspector in a munitions factory. Amy, however, thought over all the options carefully and on 20 May 1940 joined the ATA at Hatfield as a Second Officer.

The biggest hurdle that Amy now had to face was her pride. She was faced with the prospect of starting at the bottom of an organisation in which most of the women pilots, apart from Pauline Gower, were far less experienced than she herself. This did not mean that they were not well qualified, far from it. The original eight women who had been recruited by Pauline on 1 January 1940 were the pick of the bunch amongst prominent pre-war women pilots. There were: Mona Friedlander, an ice-hockey international who held both a commercial 'B' and a navigator's licence; Gabrielle Patterson, married with a small son – the first woman to gain employment as

---

* The high-sounding title of the office belied the effectiveness of the role and Lord Halifax was quick to remind him that he was, in fact, merely 'the fifth wheel on the coach'.

a flying instructor; Margaret Fairweather, daughter of Lord Runciman and the first woman to fly a Spitfire; Rosemary Rees, a wealthy young woman who had been flying her own private aeroplane; Joan Hughes, the youngest of the team, who eventually became the only woman instructress on all types of military aircraft; Winifred Crossley, daughter of a Huntingdon doctor, and former stunt pilot with an air circus; Margaret Cunnison, who spent a great deal of her time instructing new women recruits; and Marion Wilberforce. They were a formidable team.

During her first few weeks in the ATA Amy found it difficult to settle in as just one of the 'new girls'. Initially, she boarded with Winifred Crossley and five other pilots in a rented house near the airfield. It was not an arrangement into which she settled easily. She still harboured misgivings about being thrown into what she termed a 'girls' school' atmosphere, which reminded her too much of her time at the Morison's advertising agency where, being the newcomer *and* with what was perceived to be the stigma of a university degree, she became the target of cattiness from some of the other women. This time it was not her academic achievement that made her different and therefore vulnerable – many of the other women pilots would have been more than her equal in that direction. It was the fact that she was expected to live up to the reputation of being a national heroine.

Inevitably, a pecking order did exist at Hatfield, as Lettice Curtis, the first woman to fly a four-engined bomber, described in her book, *The Forgotten Pilots*. 'At times the women's pool was not unlike being back at school,' she said, and described how she encountered three well-defined strata. There were the 'seniors', mostly older women who had completed the Central Flying School course, which qualified them to fly advanced twin-engined trainers such as the Airspeed Oxford and also meant that they were trusted with the lives of up to eight of their fellow ferry pilots as they flew the Anson taxi, collecting or dropping off their colleagues at various airfields. Then there were those who were only allowed to fly the light aeroplanes used on taxi work, carrying only one or two passengers. Lastly, there were the new girls, who were only allowed to fly the Tiger Moths alone from manufacturers' airfields to Aircraft Storage Units, their instructions coming from the other two groups.

Amy went through the formality of being tested in the de Havilland Queen Bee, a radio-controlled target version of the Tiger Moth now converted back to normal pilotage and allocated to the women's pool for the initial testing of new recruits. The test was a simple, rudimentary check on the pilot's ability to navigate a short cross-country flight, to side-slip the aircraft and to recover from a spin. No doubt this must have irked Amy, who was probably doing the same manoeuvres at Stag Lane long before the person testing her had ever sat in an aeroplane.* Nevertheless, she swallowed her pride and played to the rules, knowing that because of who she was, she was most probably under close scrutiny for any sign of assumed superiority.

Having passed the first hurdle, she was accepted as a Second Officer in the ATA and allowed to wear the dark navy-blue uniform. It comprised a jaunty forage cap, military-type tunic with four large pockets, belt and brass buckle, RAF-type shirt and black tie, slacks (only to be worn on station – skirts when off duty), and pilot's wings in gold on the tunic with one broad and one narrow gold band on its sleeve and epaulettes to denote her rank. Some of the wealthier women pilots would have red satin linings sewn inside their tunics, a practice to which Headquarters turned a blind eye.

The idea of women flying in time of war evoked two quite different responses when Pauline recruited her original team of eight. The press tended to glamorise them much more than they did the men, whilst others, such as the editors and staff of some of the aviation journals, took an incredibly jaundiced view. The editor of *The Aeroplane*, C. G. Grey, a constant critic of the Mollisons during their record-making days and a man who had a reputation for going out of his way to pick a fight, made himself look rather foolish when he wrote:

> We quite agree with her [Lady Bailey] that there are millions of women in the country who could do useful jobs in war. But the trouble is that so many of them insist on wanting to do jobs which they are quite incapable of doing. The menace is the woman who thinks that she ought to be flying a high-speed bomber when she really has not the intelligence to scrub the floor of a hospital properly, or who wants to nose round as an Air Raid Warden and yet can't cook her husband's dinner.[38]

Amy's first involvement in ferrying for the ATA was flying Tiger Moths and Magisters from the manufacturers' works at Hatfield, Woodley and Cowley, to the training squadrons and to storage units in the north of England and Scotland. Five or six women pilots would normally fly their machines line-astern in a loose formation, in what they termed 'a gaggle', to airfields at Blackpool, Silloth, Kirkbride and as far north as Kinloss and Lossiemouth. Because they were in open-cockpit machines for long periods, very often in sub-zero temperatures, stops were made along the way, not only to refuel but for the pilots to thaw out their frozen limbs.

---

* Amy had 2285 hours in her flying log when she joined the ATA.

During the early part of the war many of the RAF stations did not provide facilities for unexpected women visitors, and this often presented a problem. Use of the lavatory was fraught. It was not too bad if two women were together, since one could wait outside the men's and keep watch, but if a woman were on her own it could be difficult. Alison King, in her book *Golden Wings*, gave an amusing account of how the Polish woman pilot, Anna Leska, was forced to relieve herself behind the wall of the hangar, and her wry, ungrammatical comment on the incident was: 'If the King of England himself had walked by I couldn't have done nothing about it!'

One of the worst parts of those open-cockpit deliveries to the north, which could mean being away for as long as five days at a time, was the return journey. It usually meant sitting on one's parachute in the draughty corridor of a night train back to London, and then a further train back to Hatfield. Amy took it all in her stride. In a letter home she described her new lifestyle:

> I've had a terribly busy week . . . two trips up to Scotland. It is very tiring ferrying a slow aircraft all the way and then getting home again as best we can, including a night on the train, but actually, I don't mind it very much as we manage to get some fun out of it.

The letter does show that Amy's impressions of the ATA had improved and that she now felt herself to be accepted amongst the pilots of the women's pool at Hatfield.

Amy had for many years been a very close friend of John and Alice Hofer, who owned a small precision engineering firm at Wooburn Green, near Marlow, in which Amy was a shareholder. When they discovered that she was with the ATA at Hatfield, they invited her to share their home and have her own room there. Amy jumped at the offer and moved out of her communal lodgings near the airfield after only a few months. The move had one major disadvantage, however, since it now meant that she had to make a 30-mile car journey to Hatfield every morning. It says much for Pauline Gower's kindness and understanding that she suggested an arrangement whereby Amy only had to drive the few miles to the Headquarters at White Waltham every morning, park her car and use the Fox Moth taxi aircraft to fly to Hatfield and back. How the other members of the women's pool, most of whom lived on site, felt about this privilege we do not know.

Although Amy had been flying twin-engined aircraft, such as the Dragon and the Comet during her record-breaking days with Jim Mollison, she was still required to pass the Central Flying School's conversion course at Upavon. A pass meant that she would be allowed to fly more powerful military trainers, including high-speed monoplanes such as the Miles Master and the North American Harvard, as well as the twin-engined Airspeed Oxford.

Amy may well have had some misgivings about attending the CFS course in June, for a few weeks earlier her colleague from her Portsmouth days, Joy Davison, also a pilot with PS & IOW Aviation, had been killed with her instructor when their Miles Master spun into the ground whilst making a circuit of the airfield. Two reasons have been suggested for the accident: a structural failure of the aircraft, and carbon monoxide fumes leaking into the cabin. Joy was the first woman pilot to be killed in the ATA, although technically speaking she had not been formally accepted into the organisation at the time of her death. The tragedy must have left a powerful impression upon Amy, for Alison King recounted a conversation that she had had with her which suggests that she had a premonition of her own death.

> One day I flew with her over to our HQ airfield at White Waltham – she to do a job of importance, I as a passenger to fetch some equipment. My memory of that afternoon is of easy conversation when she talked quite naturally about herself – a definite kindness on her part in that when we got to the Mess she made sure I was not left out by the crowding of her friends, and those who would be so.
>
> The talk, after much else, came on to the death the previous day of one of the pilots we all knew. The cause was probably bad weather and it was another of those tragic deaths, leaving a widow and three children. We left the Mess for our return trip immediately after that, and as we did Amy turned to me with her quick, brilliant smile, looked me full in the face with her startlingly blue eyes and said half lightly, 'You know, that'll be me one day.' Then she recoiled, as if embarrassed by this lapse and as quickly she was off down the corridor and I had to run to catch her up. Our journey home was light-hearted and neither of us mentioned what she had said. [39]

Amy passed her CFS course, having mastered the intricacies of constant speed propellers, cooling gills and boost gauges which were standard on the Master and Harvard – devices which were new to her – and landing at much higher speeds than she had been used to. Then came dual on the Oxford under Squadron Leader Bolt, which according to Lettice Curtis would have included spinning and flying on a single engine, before she was allowed to fly solo.

The Harvard was considered by some to be a flying death-trap. It was a brute of an aeroplane with a rasping, ear-splitting Pratt & Whitney engine which had certain quirks of its own. Pilots were warned by their instructors that 'this aeroplane has an unpleasant and vicious stall', and then, as if an afterthought, would come the whispered comment, 'But then of course, you don't let it stall!'[40] Some found, much to their dismay, that one could operate its retractable undercarriage inadvertently whilst still on the runway, so that as soon as the machine moved, it collapsed. This was because levers for the flaps and undercarriage were adjacent, and therefore it was quite easy for the unwary to select 'undercarriage up' instead of 'flaps up'. It did not endear the culprit very much to the ground crews.

With two broad gold bands proudly displayed on her sleeve, Amy, now promoted to the rank of First Officer with effect from 1 July, looked forward to flying not only training aircraft but also operational types such as the Hurricane and Spitfire. Although the prospect of flying fighters was a regular topic of conversation amongst the women's pool at Hatfield during Amy's time, it did not materialise until the latter part of 1941. There was keen competition amongst all ATA pilots, both men and women, to fly as many different types of aircraft as possible, but Amy and her colleagues had to be content with ferrying Airspeed Oxfords, Masters and Harvards, and occasionally flying the taxi Avro Anson on the daily routine of dropping off and picking up pilots.

As in her schooldays when she led the revolt against the wearing of 'boaters' at the Boulevard, the fact that the women pilots were paid 20 per cent less than the men for doing the same kind of work rankled with Amy. In a letter home she wrote, 'What a curious anomaly, when there's all this terrific prejudice against women even ferrying Hurricanes without guns to aerodromes for storage. I'm trying to work up an agitation about the women getting equal pay for equal work.'

Amy's concerns were not only for herself however, for with Britain now teetering on the edge of an enemy invasion after the Dunkirk evacuation, she was urging her parents, Mollie and Mollie's daughter Susan to leave the country and stay with Uncle Tom in Vancouver. Her other sister, Betty, was currently engaged to Ronald Falconar Stewart, a young Scot who had enlisted with the Green Howards at Richmond in Yorkshire, and in order to show her solidarity with her fiancé, she had enlisted as a nurse with the Anglo-French Ambulance Corps. When France capitulated, her enlistment became redundant and Amy persuaded her to join the ATA, whereupon Betty took a post which was loosely termed 'secretarial' and commenced work in the office at Hatfield in the early August.

At the start of the Battle of Britain on 10 July, when the *Luftwaffe* began its all-out assault on the airfields of southern England and Channel shipping, it was necessary for the ATA to protect its pilots, and never more so than when as many as eight or nine experienced pilots were being carried in an

Betty Margaret (née Johnson) Falconar
Stewart in ATA uniform, 1940.
(The Amy Johnson Estate)

Squadron Leader Ronald George
Falconar Stewart, DFC, was killed in
action over Berlin whilst piloting a
Lancaster of 156 Squadron during the
early hours of Sunday morning, 2
January 1944. (Grizelda (née Stewart)
Chlebowska)

unarmed Anson.

Affectionately known as 'Faithful Annie',* the Avro Anson was developed from a pre-war, six-passenger commercial aircraft design and had provision for fitting a manually operated dorsal gun turret. In spite of being a near obsolete aircraft at the outbreak of war, it was still capable of giving a good account of itself in a scrape. It had, in fact, made the first successful attack on an enemy submarine when being used by RAF Coastal Command; and in June 1940, when three RAF Ansons were attacked by nine enemy fighters, they not only survived the encounter but also succeeded in shooting down one of their attackers. With this reputation behind it, the ATA recruited twenty air gunners in the summer of 1940, a move which prompted Amy to write home, 'We're having guns in the Ansons and I have my own special gunner – a man!'

* While being used by the ATA the Ansons flew almost 10 million miles during wartime without fatal accident or mechanical failure.

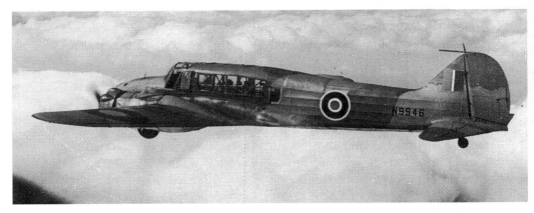

Avro Anson Mk 1, N9946, was delivered to the ATA early in 1940 and was used as a taxi aircraft. (Bruce Robertson)

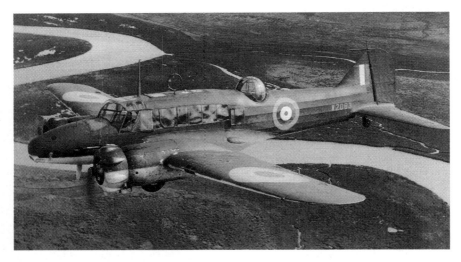

A wartime turreted Avro Anson Mk 1 as used by the ATA for taxi purposes during 1940. (R.T. Jackson)

In the summer of 1939 Jim Mollison returned to Britain, leaving his new wife in Jamaica. The marriage was not working out. His heavy drinking and womanising had continued unabated, and Phyllis perhaps expected too much from a man with such notoriously itchy feet. To sit and vegetate on a Caribbean island, albeit one in constant sunshine, was not for Mollison. He

missed the nightlife of his old Mayfair playground too much and the excitement of the café society that went with it. Knowing that a European war was imminent, it is possible that patriotism might also have played its part in his urge to return, since on one occasion he had punched a man senseless in a Bermuda nightclub for not standing during the playing of the national anthem.

When Jim drove through the gates of the White Waltham airfield to join the ATA on 1 October 1940, he had 5370 valuable flying hours in his logbook. He was the kind of man for whom they were looking, too old to go back into the RAF (he was thirty-five) even if they wanted him, and yet with a great deal of experience behind him. There is no doubt that the 'playboy airman' was an object of much curiosity and speculation amongst those with whom he was about to work. Some thought he would not last long with the outfit, since they knew he had a reputation for wild behaviour.

Strangely enough, Jim Mollison was to become a changed man, since he knew quite well that the Commander of the ATA, Gerard d'Erlanger, would not tolerate his drinking habits. Hugh Bergel, a former ATA ferry pilot and wartime colleague, had this to say about him:

> He was an infinitely nicer man than seemed possible after all the things that I had read and heard about him. As a man, Jim turned out to be delightful company. He was one of the very few people I have ever met who was completely honest about himself. He had no illusions at all about the way he had carried on and the sort of four-letter person he had been at the height of his notoriety. He became one of the ATA's most disciplined and reliable pilots . . .[41]

The main topic of conversation when Jim Mollison first arrived in the ATA, and the question on everyone's lips, must have been 'What will happen when he meets Amy?' Everyone knew that she drove her car into the White Waltham station every morning from Marlow to fly to the women's pool at Hatfield. If

Officer Commanding the ATA, Commander Gerard d'Erlanger, CBE. (ATA Association)

Amy's letters are anything to go by then he must have made the first approach, because she told her parents:

> Jim has joined the ATA and is at White Waltham! It's a nuisance, and he always makes a beeline for me and keeps on asking me out, and on these days of bad weather when we're hanging about all day it's impossible to escape him. However, it will sort itself out, and it surely must be obvious to him I'm no longer interested. Honestly, he bores me stiff!!!

One wonders if she 'protesteth too much'. Had all the old magic really gone? Many think that it had not.

Not long after Jim joined the ATA, the *Sunday Express* gossip-columnist, Lord Castlerosse, attempted to play the role of matchmaker by reporting that Amy had invited Jim back to tea at Northcroft, the home of the Hofers at Wooburn Green where she was staying, and that they had spent the evening together reminiscing and playing some of their old favourite records. It could have been a fairy-tale ending to a sad romance, for both had matured since their early days, but knowing that her ex-husband was still married to Phyllis, it is doubtful whether Amy would have risked any deep involvement– they just remained good friends.

Whilst Gerard d' Erlanger boasted that his organisation worked on an 'anywhere to anywhere' basis – an apt epithet for its abbreviation, ATA, he also firmly adhered to the dictum 'You are paid to be safe, not brave.' This emphasis on safety was enshrined in standing orders, with clear restrictions placed on bad weather flying. At the place of departure the cloud base had to be no lower than 800 feet and horizontal visibility had to be at least 2000 yards. As most of the aircraft being ferried were straight from the manufacturers and had yet to be fitted with radio, it was forbidden for pilots to fly above cloud; they had to maintain visual contact with the ground and fly by using the ¼-inch Ordnance Survey strip maps with which they were issued.

There was inevitably a certain degree of covert competition amongst pilots, with one pressing on in spite of the weather, whilst another would either put down at a convenient airfield *en route* or turn back. This could result in damaged pride for the one who played it safe. Former ATA pilot Lettice Curtis remembers that on at least one occasion Amy worried for days after turning back, even though Headquarters would have commended her judgment. Amy's flying log for that period records that she returned to her airfield of departure on two occasions; first when she was flying an Oxford from Silloth to Blackpool, and again when she was flying the Anson taxi from Llandow. It appeared that past fame could be a burdensome thing.

Whilst bad weather was the major hazard for ferry pilots, it was by no means the only one. Most large cities such as Birmingham, Liverpool, Manchester and their industrial conurbations were surrounded by a complex area of balloon barrage protection. Even in fine weather, when a pilot could follow a natural landmark such as a river or a railway line toward his or her destination, they would cause a diversion of many miles in order to avoid being entangled with balloons.

The airfields were also protected by balloons, which meant that pilots had to navigate through a lane formed by them in order to make a delivery, and in bad weather these silver-grey monsters could prove to be a fatal hazard, since mist could conceal their cables. Life was not made any easier by the fact that pilots were only allowed to look at the master map when visiting the Maps & Signals Office. These classified maps, which were kept under lock and key overnight, displayed the position of balloons and anti-aircraft gun sites to avoid. It was then a case of memorising the position of balloons and other hazards along the route one intended to fly that day (it was forbidden to mark these sites on the strip-maps), before being given a weather report and the 'colours of the day'.

Not all the hazards occurred in the air. After meeting stiff resistance from RAF Fighter Command during the Battle of Britain, the *Luftwaffe* switched its attacks from the radar stations and airfields of the southern counties and began to attack London. Early in October a Ju 88 sneaked in under the radar and made a low-level daylight attack on the airfield at Hatfield. As members of the women's pool ran to the air-raid shelters, some 50 yards away from the offices, the raider released a stick of bombs. One fell close to Pauline Gower as they ran and, but for the fact that it failed to explode, she would most certainly have been killed. Others were not so fortunate, for one of the bombs hit a factory workshop nearby, killing twenty-one and injuring seventy. The airfield's anti-aircraft gunners sent up a furious barrage of fire which made sure that four members of Kesselring's *Luftflotte* II would not live to return to base that day. The Junkers crashed nearby.

Amy's letters to her mother around this time reveal that she too was enduring night raids which deprived her of sleep. She also warned her father to avoid further visits to London. In one letter she reported that she had taken her uniform to a tailor in Savile Row for alterations, only to discover a little later that it had been completely destroyed when the shop received a direct hit. As Will Johnson's diaries reveal, there was a great bond

Senior Commander Pauline Gower and Hon. Flight Captain Jackie Cochran in 1942. (ATA Association)

of affection between him and his daughters and it was his custom to meet up with Amy and Betty whilst on business in the capital for AJK. On these occasions he would take them to the Hungaria or the Trocodero for a meal together, and they would catch up on all the family gossip. These visits became less and less frequent however, as the bombing increased.

At the end of October, Betty was given leave from the ATA and surprised everyone on her return to Hatfield by announcing that she and her Scottish fiancé, Ronald, had married. It was a typical wartime wedding, with Betty travelling down alone to his Green Howards' unit at Dorchester without the support of family or friends. They were married in a registry office on Monday, 4 November, and according to Defence Records her husband went absent without leave two days later. Evidently he had nurtured a strong ambition to become a pilot and had tried to transfer to the RAF, but his request had been met with the usual red tape from the Army. How instrumental Amy was in encouraging him to pursue his aim is uncertain, but we do know from one of his wartime colleagues, a bomb aimer who subsequently flew on missions with him over Germany, that on more than one occasion he voiced his admiration for his famous sister-in-law. Such was his determination to achieve the coveted winged brevet, that he took the law into his own hands and within a fortnight of absconding, enlisted with the RAF Volunteer Reserve. A few months later the newly-weds were separated when he was posted to Bulawayo, in Southern Rhodesia, where he commenced his flying training.

Throughout November and December Amy was busy ferrying Airspeed Oxfords, mainly to Prestwick in Ayrshire. The site for the airfield had been chosen primarily because of its clear weather conditions, since it was generally recognised that there was a sharp deterioration in weather between it and the more northerly Scottish airfields. Prestwick housed a large fleet of training aircraft and an air navigation school, as well as being the home of no. 4 Ferry Pilots Pool of the ATA.

The flight from Hatfield to Prestwick in an Oxford normally took two and a half hours and pilots followed a route which took them well to the west of Birmingham in order to avoid the balloons and the 900 foot masts at Droitwich. If the weather was bad they would put down at RAF Ternhill, just north of Shrewsbury, otherwise they would press on to cross the River Mersey in a corridor lying between Liverpool and Widnes, making sure to avoid the ships because their gunners had a reputation for being a trifle light on the trigger.

Once over the Fylde coast and within sight of the landmark of the Blackpool Tower, pilots could usually breathe a sigh of relief. From there on it was relatively straightforward as they proceeded across Morecambe Bay to avoid the balloons at Barrow. They would then fly along the Lakeland coastline to Silloth before crossing the Solway Firth to Dumfries and following the road and railway line through the Nith Valley towards the Firth of Clyde and their destination. One big disadvantage in the visual-contact flying to which ATA pilots were restricted was that with the outbreak of war all the station names had been removed in case of an enemy invasion. Before the war, pilots could navigate by following a railway line, a practice usually known as 'flying by Bradshaw', and if they were uncertain of their position they could always swoop in low over a station to get a navigational fix, but that option was now gone.

Amy had hoped to spend Christmas 1940 with her parents at Bridlington, but owing to a spell of bad weather she found herself stranded, or what she termed 'stuck out', at Prestwick for six days. On the rare occasions when she was held up there, she would stay at the Orangefield Hotel, an eighteenth-century house which had been converted into a modern hotel.* It was situated alongside the grass airfield, which at that time bordered the road running from the village of Monkton to Ayr, and its management prided itself in providing a swimming pool and squash courts for its guests. No doubt these facilities helped to soften the blow somewhat for Amy during her enforced stay there.

Also staying at the Orangefield during that holiday period was the young fighter pilot Sandy Johnstone, DFC, who was serving with 606 (City of Glasgow) Squadron based at Prestwick. In his diary he recalled that after a shopping expedition in Ayr, 'Amy Johnson was at the hotel when we got back, having flown in to pick up an Oxford Trainer being refurbished by Scottish Aviation Ltd, but it is not ready for collection and we have invited her to join our party.' A few days later he wrote:

> Prestwick was blanketed in fog throughout Christmas Day, an unusual occurrence in this part of Scotland . . . thereby scuppering Amy's hopes of getting away and condemning her to a spell of Scottish rowdyism instead of the hoped-for sophistication of an English yule-tide occasion. Nevertheless she seemed to enjoy herself amongst strangers and certainly succeeded in enrapturing Dad when he found himself sitting next to her at dinner.[42]

On the 28th he saw her off in her Oxford in time to be back at Hatfield for the New Year celebrations.

On the morning of Friday, 3 January, Amy was detailed to ferry yet another Airspeed Oxford to Scotland, again to Prestwick, and to collect a similar machine which was needed at Kidlington, near Oxford. Her colleagues knew that she was hoping to be back at Hatfield on the following Monday for a party which was planned to celebrate the first anniversary of the formation of the Women's Pool at Hatfield. The weather was notoriously bad during the winter of 1940/1, and Amy met with poor flying conditions in the Midlands on the way north. She was forced to put down at Ternhill and spent the night at the Hawkestone Park Hotel in Weston-under-Redcastle.

The following day Pauline Gower took a phone call in the Operations Room at Hatfield to learn that Amy had arrived safely at Prestwick, and that she had left late in the afternoon for Kidlington, saying that she might put down at Squires Gate if necessary. Now it seems as if Amy was deliberately taking the opportunity to visit her sister Mollie in Blackpool. She needed

---

* This well-known hotel was demolished in 1966 to make way for the southerly perimeter track accessing two of the runways of what is now Prestwick International Airport.

little excuse for doing so, since ATA pilots were forbidden to fly after dusk.

When Trevor and Mollie Jones returned home from a shopping expedition that Saturday afternoon, much to their surprise they found Amy sitting in their lounge waiting to greet them. Amy explained that flying conditions had been bad and that she suspected that her compass was slightly inaccurate. Whether this was true or not is debatable; it seems unlikely, as there is no record that this was ever reported when she landed at the Squires Gate aerodrome, and when Trevor offered to phone the airfield and request that it be looked at, Amy stopped him, saying that she did not want to be delayed from taking off the following morning. No doubt she was anxious not to miss the celebration that would be taking place at Hatfield on the Monday. The matter was then dropped and they spent a pleasant evening together chatting over family news.

Chapter Twenty

NEVER ON A SUNDAY

cold, grey sea-mist hung over the coastal aerodrome at RAF Squires Gate, Blackpool, on the morning of Sunday, 5 January 1941. Dispersed around its perimeter one could just discern the ghostlike shapes of a few Ansons and Bothas of the School of General Reconnaissance, whilst the four Hurricanes of 96 Squadron stood like silent sentinels in front of the hangars. Just before 8 a.m. the civilian members of the ground staff of the Brooklands Aviation Company began to arrive, amongst them the fitters and riggers of the duty flight. The first item on their worksheet for the day was to prepare Airspeed Oxford V3540 for Miss Johnson's departure to Kidlington, a flight which was scheduled for 10.45 a.m.

Meanwhile, less than 4 miles away, the bedroom curtains at 143 Newton Drive were being drawn aside and Mollie Jones was warning her sister, 'Amy, you can't possibly fly today, the weather is simply awful!' Amy stretched and rubbed her eyes, 'Oh, but I must. I'll be all right. The plane's needed and I do want to make the party at Hatfield on Monday.' After breakfast Amy began packing her bags, and thanking her sister once again for the two belated Christmas presents that she had been given: a set of satin underwear from Mollie and Trevor and a large oval mirror edged with a fresco of flowers from her two-year-old niece, Susan. By 10 a.m. Amy was ready to leave, and Mollie stood at the front lounge window holding her young daughter, as they waved goodbye to their guest. It was the last time that they were ever to see Amy.

Amy Johnson was always a popular figure amongst the people she met at wartime airfields, and with whom she would enjoy a laugh and a chat, especially those who were servicing her aircraft. Harry Fidler was the

Brooklands' Aircraft Rigger, Harry Fidler, 1940. (Harry Fidler)

twenty-one-year-old rigger who checked the controls on Amy's machine that morning, and he recalled his colleagues asking her if she had enjoyed a good Christmas, whereupon she joked that she had even received a pair of slippers from her ex-husband, Jim Mollison.

Also at the Squires Gate aerodrome that morning were two other ATA pilots, Captain Joe Shoesmith and First Officer Greg Piddocke, both from the ferry pool at Ringway. They were due to deliver Spitfires back to their base, but upon entering the Watch Office they were advised to wait for an improvement in the weather and if necessary to stay over for another night. They needed little persuading, but Amy was already in the cockpit of the Oxford and, it seemed to them, was determined to go. It was left to the discretion of each ATA pilot whether they took off or not, but there were strict rules on the required visibility for flying and on this particular morning they were not met.

The petrol bowser driver, Harry Banks, topped up the Oxford's two auxiliary tanks with 32 gallons of fuel, which meant that the aircraft was now carrying its maximum capacity of 156 gallons, enough to give the aircraft a duration of at least five hours' flying time. It was more than ample for the short flight its pilot was about to make.

Between 10.30 and 11.30 a.m. an intermittent drizzle began to set in, but in spite of this Amy asked Harry Banks to connect the starter battery, whereupon she fired up both the aircraft's engines. With the engines ticking over, Banks then climbed into the aircraft to join Amy in the cockpit, where she offered him a cigarette and they sat chatting and smoking whilst she waited to see if the weather improved.

Around this time, Greg Piddocke recalled:

> Captain Shoesmith offered to take the latest information out to the waiting Oxford. I accompanied him on the wet dash to the aeroplane, we

signalled her to come aft from the cockpit and Amy came to the door. Buffeted by the slipstream from the prop and thoroughly soaked by the rain, Captain Shoesmith delivered the latest actual weather to her. She spoke briefly to my friend, then closed the door. While we departed swiftly for shelter, the aeroplane taxied out and took off.[43]

Amy had decided that she would 'go over the top', meaning that she would fly above the cloud layer and trust to finding a hole in the clouds somewhere near her destination. It was exactly 11.49 a.m. by local time when the duty pilot noted her time of departure.

As Amy climbed to approximately 6000 feet to fly southwards, she found herself above a marbled carpet of grey cloud with blue skies and brilliant sunshine above her. With an indicated air speed of 120 m.p.h., she would then have moved the mixture control lever to the 'weak' setting in order to minimise her fuel consumption. The surface wind was in a north-easterly direction at 15 m.p.h. and the airstream was therefore assisting her. What she did not know was that a layer of unbroken stratus, its base varying between 800 and 2000 feet, covered the whole area of the Midlands and the south of England into which she was now flying.

Kidlington was a mere 150 miles from Squires Gate, which meant that she would have been over her destination at approximately 1.05 p.m. and by then vainly looking for a hole in the clouds through which to descend. If she did attempt to break through the cloud layer, she would have been confronted with two problems. First, she could not be sure of her exact position, since she was flying purely by dead-reckoning, that is flying by the compass alone (no radio or radar assistance) and estimating her drift. Although the airfield at Kidlington is only 80 feet above sea level, there was always the possibility that she would fly into higher ground unexpectedly if she did decide to chance it and descend through the cloud layer. Secondly, there was what is known in meteorological terms as an inversion on that day. This means that a layer of very cold air near the earth's surface is trapped beneath an overlying layer of warmer air. If she did attempt a descent, she would have met with icing conditions which would cause ice crystals to form on the windscreen, making it difficult for her to see.

Amy was now left with three options: make a reciprocal compass heading and return to Squires Gate – a blow to her pride; proceed in a south-easterly direction and look for clearer weather around airfields with which she was familiar; or circle the presumed site of the Kidlington airfield in widening circles and hope for the weather to improve, or for a hole to appear in the cloud through which she might make visual ground contact. There is reason to believe that she took the last course initially. It was a situation similar to that which she and Jim Mollison had experienced when they became lost in *Black Magic* during the 1934 MacRobertson Air Race. On that occasion, they were forced to circle an unknown town in India (it turned out to be Jabalpur) in the darkness for some considerable time until there was sufficient daylight in which to make a safe landing. It is quite feasible that Amy could have circled her Kidlington destination for perhaps an hour, by which time it would be approximately 2.05 p.m.

Having searched in vain for a break in the clouds, and perhaps having attempted a descent into the clouds on more than one occasion, she would have reasoned that the best plan of action was to set course for her home base at Hatfield. If that was under a similar cloud layer, she would switch onto a 120-degree heading and proceed towards the North Sea. Once sufficient flying time had elapsed she would have been sure that she was flying over water, with the assurance that there was no possibility of flying into dangerously high ground when she did finally descend.

Most experienced pilots confirm that this is what they would do under similar circumstances, and it is fairly certain that this is what Amy did. According to Pauline Gower the Oxford was tracked by radar from one of the Chain Home Low stations, either at Foreness, near Margate, or at Walton-on-the-Naze. She was told that RAF fighters had been alerted from Hornchurch during the afternoon but failed to intercept. These CHL stations were quite capable of picking up an incoming aircraft as low as 800 feet and as distant as 20 miles.

Amy would still have to deal with the problem of icing when she finally did descend, but if the descent was made at a very shallow angle, she would have broken cloud at a safe height of somewhere between 600 and 800 feet. This was the height of the cloud base for that area of the UK on that particular day according to the Met Office report. Once below the cloud she would have turned on a reciprocal heading by turning through 180 degrees, making visual contact with the sea below through the clear-vision side panel and by opening her side window. With the cockpit heating fully on, any windscreen icing would begin to clear. Amy would now be free to head back to the Essex coast, perhaps not an easy task if she was flying into the sleet and snow showers that were falling along the Thames estuary on that afternoon. The time would now be somewhere between 2.50 and 3 p.m. and she would already have switched from the main fuel tanks to the auxiliaries. However, she would have still had at least a further ninety minutes' flying time left.

Amy was familiar with most of the airfields in that area of the Thames estuary from her pre-war flying days, and it is highly likely that Southend's aerodrome at Rochford would be uppermost in her mind. If Muriel Hollington is to be believed (see chapter 1), and we have no reason to doubt her, then Amy was probably searching for this airfield when she flew over the young woman's house that afternoon. The aircraft was at rooftop height as it headed out back into the estuary, and was far too low for its pilot to bale out.

The Balloon Barrage Vessel, HMS Haslemere, *in wartime patterned camouflage.* (Maritime Photo Library)

What we know for certain is that Amy Johnson's Airspeed Oxford was seen to make a landing in the sea near the East Knock John buoy and alongside one of the ships escorting a convoy, HMS *Haslemere*, commanded by Lt Cdr Walter Edmund Fletcher, RNR. Convoy East 21 comprised seventeen merchant vessels escorted by eighteen naval ships: the two Hunt Class destroyers *Fernie* and *Berkeley*, two anti-submarine trawlers; a Chasseur (a French submarine chaser); four minesweepers, flying very low altitude Kite balloons; four motor launches; and five barrage balloon vessels: *Haslemere, Pintade, Pingouin, Roebuck* and *Sambur*, each flying the LZ, low zone balloons normally used around land-based defences.

According to Temporary Lieutenant Henry Patrick O'Dea, RNR, who was on the bridge of the *Haslemere* alongside Fletcher at the time, he had seen a parachutist descending out of the clouds with an aircraft circling the person coming down at about 3.30 p.m. He then saw that 'the aircraft glided down, *probably under some measure of control*, but with its engines stopped', and that it '*broke up at once.*' The *Haslemere* immediately altered course to starboard in order to effect a rescue of what were thought to be two survivors, but at this point the ship ran aground in shallow water. The ship's

The Type 1 Hunt Class destroyer, HMS Garth, *sister ship to the* Berkeley *and the* Fernie *which were escorting Convoy CE 21 on 5 January 1941. These small warships packed a powerful punch when approached by unidentified aircraft, having quadruple pom-poms and two twin 4 inch HA/LA anti-aircraft guns, one forward and one aft.* (Maritime Photo Library)

Lieutenant Commander Walter Fletcher, RNR, with fellow officer Lt. McKinley, RNR, on the bridge of HMS Haslemere. *Fletcher lost his life shortly after this picture was taken.* (Shann Puckle)

engines were immediately put into reverse, but it took more than ten minutes before it was free of the sandbank. By this time one of the 'survivors' was about 40 yards from the wreckage, whilst another, thought to be a woman, was about 20 yards from the stern of the ship and is reported by Leading Seaman Nicholas Roberts to have called out, 'Hurry, please hurry.'

Two heaving lines were thrown to the woman by members of the crew, but she did not manage to grab either of them. At that point Seaman Raymond Dean went over the side of the ship and down onto the rubbing strake, a platform which runs around the side of the vessel a few feet above the water line. Lying prone on it and with one arm through a fairlead to secure himself to the ship, he stretched out with the other to reach the woman, who was almost within his grasp. The stationary ship was heaving with the strong tide, causing the stern to rise and fall in the heavy swell. The last Raymond Dean saw of the woman was as the stern came crashing down and she disappeared beneath it.

Meanwhile, believing that the woman at the rear of his ship was near to being rescued, the *Haslemere*'s captain ran from the bridge, took off his duffle-coat and boots and dived (some witnesses say he jumped) into the freezing water to swim towards 'the other survivor'. By this time the ML 113 and the *Berkeley* were closing in on the scene of the crash, because they could see that the the crew of the *Haslemere*'s lifeboat, with seven of its men on the oars, was having difficulty making any headway at all in the heavy seas. Walter Fletcher was seen to be supporting what was thought to be a man in the water for up to three minutes, before relinquishing his hold and swimming back towards a Carley float which had been launched by ML 113. When he was eventually pulled from the sea he was unconscious and, in spite of an attempted resuscitation, he died shortly afterwards from hypothermia and shock.

As soon as this version of events was given out in an Admiralty's press release a few days later, all manner of wild stories began to circulate. The most lurid was that Amy was on a secret spying mission and that she was taking an agent to France when the accident occurred, or that she was smuggling a German friend out of the country. The newspapers wanted to know, 'Who was this passenger, Mr X, the man who accompanied Amy Johnson on the day of her death?' Naturally, these quite unfounded rumours caused considerable pain to the Johnson family and infuriated members of the Air Transport Auxiliary.

We now need to look carefully at all the facts if we are ever to dispel some of the popular myths and get to the truth in the light of new evidence by examining what actually happened on that Sunday afternoon of 5 January

1941. As the Queen of Sheba once said to King Solomon, 'Indeed, the half has not yet been told.'

The rumour that Amy was on a secret spying mission for the Government at the time of her death probably arose because of Amy's appearances at various RAF stations in France during the 'phoney war' period of late 1939 and early 1940, when she was flying VIPs and equipment for National Air Communications. Added force was given to the rumour by her meeting with Lord Vansittart, to whom she offered her services just before the Special Operations Executive was formed. Unfortunately, one of Amy's colleagues, Alison King, who was the Operations Officer of the Hatfield Ferry Pool at the time of the incident, perpetuated this myth when she wrote an article for *Aeroplane Monthly* in December 1980, in which she stated that she was 'not convinced that Amy was flying alone when she came down.'

If she had been secretly chosen by SOE to fly an agent into or out of occupied France, then the Airspeed Oxford, with its undersides painted yellow to show that it was a training aircraft, was the least likely aeroplane to be used. For a parachutist to jump from it meant jettisoning the door, an action that would have meant the pilot flying back to this country with a gaping hole in the fuselage that would cause the aircraft to be buffeted about in the slipstream. Moreover, such a flight would most certainly not take place in daylight, or from such a northerly point of departure.

In order to prove beyond a shadow of doubt that Amy was alone when she left Squires Gate, Amy's solicitor sent his senior investigator, John Don, to the aerodrome to interview and take statements from seven members of the ground staff, most of whom were able to verify that the pilot was the only occupant when she took off. The possibility that she could have landed at another airfield to pick up a passenger after taking off from Blackpool was soon dismissed when a signal sent on 6 January to all RAF flying units received negative replies. If Amy had been delivering a Lysander, an aircraft used by the SOE later on in the war because of its short take-off and landing run, then the story might have had some credibility, but to have landed the twin-engined Oxford in a meadow or a field was simply out of the question.

The even more sensational rumour that Amy was flying an alien friend back to enemy-occupied territory was probably nothing more than speculation by those who knew of her former love affair with Hans Arregger, a German-speaking Swiss. Amy was too much of a patriot ever to betray her country, and that rumour was soon dismissed. However, when

Amy's former employer, Sir William Crocker, conducted his enquiries at the Probate Court in December 1943 at the request of the Johnson family, he had difficulty obtaining statements from naval witnesses. In a letter of 9 July 1942 to Will Johnson he wrote: 'One would be doing violence to the language to suggest that the Admiralty have been really helpful.' Crocker needed to establish the 'presumption of death' of his client in order to administer her estate, since her body was never recovered. More importantly, he wanted to avoid any stigma being attached to Amy's name – that she was acting illegally in any way, or engaged in some amorous adventure.

Lieutenant O'Dea, both in his written deposition of 19 August 1941 and in his personal appearance in court two years later, stood by his statement that he saw two people in the water after the Oxford ditched in the sea. This was also the view of Seamen Roberts and Dean who had made attempts to rescue the woman at the stern of the ship; whilst Seaman Vivian Gray, a member of the lifeboat's crew, said that the 'man' Fletcher was supporting appeared to be a German airman because of his flying helmet.

Two bags containing Amy's clothing and personal possessions, together with the Christmas presents from her sister's family, were recovered from the sea by the crew of a whaler launched from the *Berkeley*. The skilful lawyer maintained that the 'male passenger' was nothing more than one of these two bags, a zip-fastened pig-skin bag which was sufficiently airtight and waterproof to float for a short time. He argued that it was unlikely that a man as brave as Fletcher would have given up and abandoned a rescue attempt if he had been holding onto a person, and that

Amy's pig-skin bag which was mistaken for the mysterious 'Mr X'. (East Riding of Yorkshire Council)

he was merely supporting himself on the bag until either the lifeboat from the *Haslemere* or the Carley float launched from ML 113 drew nearer.

Crocker obviously felt that it would be too great a coincidence if it were the body of a dead German airman from some previous action; even if it were, it posed the question whether Fletcher would have let go of someone who still might be capable of responding to resuscitation. His selfless action in diving into an icy cold sea suggests not.

Examining the yellow pig-skin bag today in the Amy Johnson Room at Sewerby Hall, Bridlington, one can see how its rounded pyramid shape could easily have been mistaken for the helmeted head of an airman. Sadly, it seems that the brave Fletcher died in a vain attempt to save a supposed survivor, although his bravery was justifiably recognised when he was posthumously awarded the George Cross (his mother and brother attended the Buckingham Palace ceremony on 10 June 1941 to receive from King George VI what was then the Albert Medal). Those in authority took a less charitable view of Fletcher's rescue attempt when, soon after his death, the Admiralty's Director of Navigation expressed the view that 'it is wrong in principle – especially in war – for any Commanding Officer to abandon even temporarily the command of his ship, by himself diving overboard for the purpose of saving life at sea'.

We now come to the question of the parachute descent said to have been made by Amy on that Sunday afternoon. There is no doubt that she was wearing a parachute when she left Squires Gate, since this was confirmed by members of the ground staff servicing her aircraft. However, it was a necessary and routine practice for a pilot to wear one, since the metal bucket seat required either a parachute or a thick cushion to provide a comfortable seat for the occupant. Amy would then have secured herself into the seat by means of the four straps of the Sutton harness which were attached to the seat's frame. In an emergency she could release either the seat harness alone if she intended to bale out of the aircraft, or both seat and parachute harnesses in the event of a sea-ditching.

During the proceedings of the Probate Court case Lieutenant O'Dea stood by his account that 'a parachutist was seen dropping out of the clouds', and that 'an aircraft was then observed to be circling the parachutist'. Seaman Dean, who was on the flag deck of the *Haslemere* at the time, said that 'there was only one parachutist'. It is interesting to note that Crocker seemed to accept their evidence at face value, without questioning their assertion that Amy had baled out of her aircraft, his main concern being to protect Amy's 'high sense of duty and honour' by disproving that there were ever two people in the water. The doctor of the *Berkeley* subsequently maintained that he had at one time counted *eight* parachutes during the Oxford's descent (this dubious piece of evidence will be dealt with more fully in due course).

The American journalist Drew Middleton was amongst the eye-witnesses on convoy CE 21 during that afternoon and his account throws some doubt on the suggestion that a parachutist was seen to leave the Airspeed Oxford. He recorded that he had first seen Amy's plane 'at 750 feet, gliding with its engines silent; then at 200 feet something white fluttered out of it, perhaps, *but not certainly* a parachute'.

Lieutenant Ian David McLaughlan was on the bridge of the *Berkeley* when the Oxford came into view, and in his deposition he testified that he 'did not see any parachutist'. His evidence is backed up by Seaman Dennis Turberfield who was a gun-layer on the same ship during the incident, and he said, 'At no time did I observe any one or more persons, or even objects, leave the aircraft.'[44] Added weight is given to McLaughlan's and Turberfield's testimony by virtue of their ship's close proximity to where the Oxford landed in the water. The First Officer, Lieutenant Julian Pearson, who was also on the *Berkeley*, stated that 'the day was cloudy and the plane came down 100 yards from the ship', and that 'a ship's boat recovered her luggage'. Again, there was no mention of a parachute. Furthermore, the crew of ML 113 only observed the aircraft descending, they did not see a parachutist.

Neither a parachute nor a jettisoned door was ever recovered from the scene of the crash, even though both would have floated and been seen easily; the door was wooden and of hollow construction, whilst the 'chute was likely to billow above the surface of the sea, with or without a parachutist attached to it. (It was not unknown for a wartime pilot to survive a sea-ditching by floating on a ballooning parachute until rescued.)

Old habits die hard, as the saying goes. Amy was prepared to ditch near Spurn Head alongside an ocean liner in the North Sea in November 1930, rather than bale out, when she became lost in similar weather conditions to those that existed on the day of her death. Moreover, although she carried a parachute on her Australian flight, she never considered using it, even though

Muriel Hollington. (M. Hollington)

there were circumstances when she might well have been tempted to do so. On her two solo flights to the Cape and back in 1932 and 1936, she did not carry a parachute, preferring to 'stick with the aircraft even if it consists of only a half a wing'. ATA colleagues have also confirmed that her philosophy was that in an emergency she would never take to the silk, but would always try to bring the aircraft down. If she had been tempted to bale out at the dangerously low heights reported by Muriel Hollington and Drew Middleton that day, then the mental picture she retained of Florence Klingensmith's body falling with an unfurled parachute wrapped around it might quickly have deterred her from doing so.

Some have asked whether there is any possibility that, having stayed with the aircraft and made a successful landing in the sea, Amy could exit either through the roof panel directly above her (the quickest way out), or via the normal exit door at the rear. An escape through the roof only required her to pull one of two red handles, an action which automatically rips the securing tape and enables the panel to be pushed out. Alternatively, she could move to the rear door by making use of the rope handholds which run along the upper part of the fuselage. In either case she could then scramble out onto the wing of the floating aircraft. This would not be possible if Lieutenant O'Dea's assertion that the Airspeed Oxford broke up almost as soon as it landed in the sea is true. However, one is forced to question the accuracy of O'Dea's statement in the light of a letter written to Alison King by the captain of the *Berkeley*, Lt Cdr H.G. Walters. In the letter he said, 'My whaler returned having found no human being in the *still floating wreck*, but brought in a leather bag. It is simply not true to say . . . that the aircraft started to break up at once in the heavy sea.'[45]

Lt Cdr Walters's statement certainly suggests that the Oxford must have made a controlled and relatively smooth landing in the sea if it was still floating by the time his whaler reached it. An Airspeed Oxford was not the easiest of aeroplanes to fly, for it had a reputation to swing just before touchdown if not handled carefully, and that with a pilot at the controls! If it had been pilotless, it would certainly not have made the smooth landing in the sea that it did.

The captain's testimony also has a bearing upon a comment made by Pilot Officer Anthony Puckle, who was the RAF balloon officer on the *Haslemere*, for whilst being interviewed by a newspaper reporter in 1988, he stated: 'At the time there was talk of another body in the water and I must admit I thought I saw a figure standing on the wing when the aircraft was in the water. But others tell me that what I probably saw was Amy's parachute billowing behind her.'[46]

Flying Officer Anthony Puckle, the RAF Balloon Officer on HMS Haslemere *on 5 January 1941.* (Shann Puckle)

The air historian Roy Nesbit has undertaken considerable detailed research into the events of the day that Amy died, paying particular attention to the part that the defence balloons may have played in her death. He discovered that at this stage of the war a line of twelve drifters (small fishing vessels of approximately 100 tons), each carrying a balloon, were moored at regular intervals across the mouth of the Thames estuary between Southend and Sheerness. Their task was to deter enemy aircraft from carrying out accurate low-level attacks in daylight and to hinder enemy mine-laying aircraft from dropping their deadly cargoes by night. Each drifter was manned by a civilian crew, together with three or four RAF personnel to handle the balloon and its winch. One of the drifters was struck by an enemy mine and sunk with the loss of all on board on the night of 17 December 1940, and the crews of six of the remaining drifters were evacuated. Their balloons however, were left in position and these were seen and reported by the Hurricane pilots of 249 Squadron to be flying above the cloud layer nine days before Amy died. On 3 January one of the unmanned drifters was sunk by heavy seas, leaving five balloons still flying above cloud two days later.[47]

It has been claimed that Amy might well have seen this line of standard LZ balloons on that afternoon, and some people have therefore speculated that she baled out over them in the belief that she would land on terra firma. Amy would have been quite familiar with the difference in shape of the two types being flown by the convoy; the LZ is bulbous with rounded, quilted fins; whilst the sea-going Kite balloon was much sleeker and had three sharklike fins. Of course, it is possible that if she was above cloud at the time, and if she did see them, she could have descended around them to investigate her chances of making a landing. However, it is highly unlikely that she would have jumped, even if they were of the type usually flown around land-based defences.

Vessels of the Mobile Balloon Barrage Flotilla moored alongside the dock at Southampton and waiting to escort another Channel convoy. (Shann Puckle)

The idea that Amy made a parachute descent on that fateful afternoon has been further strengthened over the years by the repeated showing on television of the 1941 Herbert Wilcox film *They Flew Alone* (US title, *Wings and the Woman*), starring Anna Neagle as the aviatrix. Today, everyone takes it for granted that Amy jumped, because they have seen it on the screen. And so the myth has been perpetuated.

On the afternoon that Amy died, her ex-husband, Jim Mollison, was ferrying an unarmed Hawker Hurricane fighter on a routine delivery flight from the aircraft factory at Langley, Buckinghamshire, to the airfield at Aston Down, near Stroud. As he taxied in towards the hangars and switched off the engine, he glanced at his wristwatch to note his time of arrival. It was precisely 2.45 p.m. Little did he realise that at that very moment his former wife was probably in a cold sweat somewhere over southern England, as she searched in vain for a break in the clouds.

Jim Mollison believed to the end of his days that Amy was the victim of enemy action, shot down by a German raider. He had good reason for doing so, since he himself had been fired upon by a lone Me 110 whilst flying the taxi Anson back to White Waltham. Fortunately, the enemy did not press home his attack. On that occasion Jim had eleven of his pilot colleagues on board and their loss would have been a devastating blow to the ATA.[48] However, it is on record that a total of nine Oxfords were either damaged or shot down by enemy aircraft during 1941, most of them at night whilst they were on training flights. But in Amy's case the *Bundesarchiv* records for *Luftflotte* II, the Brussel's-based wing responsible for covering the Thames estuary sector, fail to show any aerial combat for that particular Sunday afternoon.

If Amy was not the victim of a lone *Luftwaffe* attack, one is left with the question whether she might have been the victim of 'friendly fire' from either the ships or the coastal batteries. Of the ten depositions which William Crocker gathered from naval eye-witnesses for the Probate Court, only Lieutenants O'Dea and McLaughlan make mention of gunfire, and in both cases it is to deny that there was any at the time the Oxford landed near them! O'Dea said, 'I did not hear any machine-gun or other fire at this time', and McLaughlan said, 'I heard no machine-gun or other fire at about the time I saw the aircraft'; their statements are almost uncannily identical. One is reminded of the saying that 'their stories have this in common with the Laws of the Medes and Persians, they alter not.'

There was said to have been an enemy attack on the convoy by a solitary Junkers 88, either during the time when the Oxford ditched in the sea, or before, or soon after, depending upon the witness. In a letter to Alison King, the *Berkeley's* captain stated:

> My main attention at the time was taken by a lone German bomber who

was doing his best to be a mild nuisance at my end of the convoy, but this didn't last long, and when I looked again I was surprised to see that the *Haslemere* seemed quite some distance from the *ditched* plane.[49]

It must be remembered that the convoy was now in single line ahead over a distance of 2 to 3 miles, owing to the narrowness of the channel in that part of the Thames estuary.

A signal to the Commander-in-Chief Nore Command from the captain of HMS *Fernie*, the other Hunt Class destroyer at the head of the convoy, stated that the *Berkeley* signalled to him at 4 p.m. that the rear ship of the convoy had been bombed by a Ju 88 and that the *Berkeley* had opened fire on it. Seaman Dennis Turberfield, a gun-layer on the *Berkeley*, was adamant that 'the Ju 88 skirmish followed later on the starboard quarter'. However, not all his colleagues agreed with him for in June 1988, Ernest Hannam, who had been a leading stoker on the destroyer, said: 'We were certainly not under attack after Amy Johnson's plane crashed, it came before.'[50] Earlier, in a statement to the press, he was even more forthright when he stated: 'I really believe, deep down in my heart, that we blew her out of the sky.'

A British destroyer and colliers under fire in the English Channel, August 1940. It was an attack such as this in November 1940 in which HMS Haslemere *had suffered a near hit resulting in the loss of its First Lieutenant.*
(Imperial War Museum)

One is forced at this point to ask if the records actually confirm an enemy attack on convoy CE 21 in the Thames estuary area during the afternoon of 5 January 1941. Alison King consulted the Air Ministry War Room Daily Report Summary, and whilst it showed that a number of enemy aircraft reconnoitred shipping in the area, and that bombs were dropped in East Anglia, these were in the morning and not in the afternoon.

According to Roy Nesbit, the German records show that *Luftflotte* II and III despatched about forty-five bombers and reconnaissance aircraft during daylight hours, some of which attacked 'a convoy of sixty-two ships' seen entering the Thames estuary, but with limited success.[51] It needs to be remembered that as many as 200 merchant vessels and their escorts could pass through the mouth of the estuary on any one day, and that the enemy report does not specifically state that any attacks were made on a convoy the size of CE 21 during that afternoon.

It is significant that the only report which specifically states that an attack was made on CE 21 comes from the War Diary of the Nore Command itself, which reads: 'Convoys FN 75, FN 76 and CE 21 were attacked by aircraft off Cromer, Harwich and in the Estuary respectively, but in no case was any ship hit or damaged.'

We do not know how many rounds the *Berkeley* fired on this occasion, or whether any of the other escort vessels and merchant ships also opened up on the alleged Ju 88. The normal rules of engagement would be to identify the aircraft on the radar screen to see if it was friendly. In order to do this, a device known as IFF (identification friend or foe) was fitted in the nose of Allied aircraft, which enabled them to respond automatically to a radar signal from ship or shore by sending back a more powerful signal on the same frequency. This signal would then appear as a blip on the screen at the receiving end so that they could be identified as friendly aircraft. Any aircraft approaching within 1500 yards of our shipping and not responding to IFF would be fired upon.

The pivotal question now seems to be, if the Ju 88 was fired upon by the convoy, then why should the same ship, or ships, not have fired upon an Airspeed Oxford which was not fitted with IFF? Aircraft recognition was never a strong point with naval gunners, who had a justifiable reputation for being a trifle light on the trigger finger. In wartime it was very much a case of shoot first and ask questions later. Moreover, it would not be easy to distinguish between the silhouettes of the two aircraft. Both were twin-engined and not dissimilar when viewed from a head-on position, particularly under the poor weather conditions which existed on that day. It was not unknown for Allied aircraft to fall victim to 'friendly fire' from ships.

How much harder it would be to admit to this when a national heroine was involved!

There is also the question of the possible sighting of Amy's aircraft and gunfire from coastal batteries of the Royal Artillery. Could it be that the eight 'parachutes' that were supposedly seen by Dr W.T.K. Cody of the *Berkeley* were actually white puffs of gunsmoke in the sky?

As we saw in the opening chapter, Gunner Tom Mitchell of 207 Battery of the 58th Heavy AA Regiment stationed near Iwade in Kent, claimed that their battery of four 4.5-inch guns had fired on an unidentified aircraft on that Sunday afternoon. Subsequently, upon learning that Amy's aircraft had crashed in the Thames, his officers had warned him and those manning the guns that day not to breathe a word of their part in the action to anyone. However, in spite of the warning, he had written to his sister soon after and told her of the incident. He had not gone public with the story until February 1999, when, following his sister's death, a bundle of old wartime letters that she had kept, including that one, was returned to him.[52]

Many people have sought to dismiss Tom Mitchell's account as nothing more than the ramblings of an old soldier looking for some cheap publicity. The main objection to his testimony appears to be that he claimed the aircraft was radioed by his battery to give the colours of the day and the pilot failed to respond, whereupon the guns were ordered to open fire. Critics say that his account cannot be true because the Airspeed Oxford was not fitted with a radio or IFF, therefore, according to them, the entire content of his account of the incident should be dismissed. More cold water was poured on his version of events when it was found that the official records showed that his battery did not fire any rounds until about 7 p.m. and that it ceased to fire at about 11.30 p.m. on 5 January 1941.

The author felt that if Tom Mitchell could produce the original letter then it might go some way towards substantiating his account but, he could

Gunner Tom Mitchell of the Royal Artillery. (Tom Mitchell)

The gun crew of Battery 207 of the 58th Heavy AA Regiment at Iwade, Kent, which Gunner Mitchell claims to have fired on Amy's Airspeed Oxford. (Tom Mitchell)

not do so, claiming that he had lost it. However, the newspaper reporter to whom Mitchell had first given the story is prepared to swear under oath that she saw it.[53] There is still the problem of the official records, which show a mismatch in the time of the incident and the time when the guns were reported to have opened fire. However, one must always bear in mind that records are only ever as accurate as those compiling them!

Tom Mitchell is not alone in claiming that his gun battery had contact with Amy's aircraft, even if it was not a visual one. Gunner Richard Powell, who was stationed with 419 Battery of the Royal Artillery near Shoeburyness said in 1988, 'Her plane was circling round and it was being shot at by the British ships. I could see its RAF markings but perhaps they couldn't because of the bad weather.[54] Amy's Oxford was also sighted by the Heavy AA unit stationed on the Belfairs Golf Course at Leigh-on-Sea, although there is no record that they fired at it.[55] This sighting most probably took place just moments before the Oxford flew over Muriel Hollington's house at rooftop height.

We cannot be sure of the exact time that Amy's plane landed near the ships of convoy CE 21, because of the conflicting evidence given by the naval witnesses. As we have already seen, the times given vary between 3 and 3.30 p.m., with the most precise being 3.37p.m., a time given by Lieutenant Loasby, the Senior Officer of ML 113. If we accept his time, it would mean that the Oxford was in the air for three hours and forty-eight minutes. Some have argued that the times given by the Royal Navy in their reports would be in GMT and not in BST.* If this argument is accepted, it would mean that the Oxford was in the air for four hours forty-eight minutes and almost out of fuel when it came down. However, a statement made by Seaman Joseph Henry of the *Haslemere* causes us to doubt his use of GMT. He said, 'At about 3 p.m. . . . I was in bed and called out to man a lifeboat.' It would have been natural for any man awakened from his sleep to look at his watch, or the bulkhead clock, when recalling the time of such an important incident. Furthermore, the scholar and ex-naval officer, H. D. Howse, author of *Greenwich Time* is 'pretty sure that the escort vessels kept local civil time'. The argument that Amy was almost out of fuel when she appeared over the convoy is therefore very much open to question.

Later in the afternoon of the incident, H.M. Drifter *Young Jacob* was on patrol when one of its seamen spotted a 10 foot by 4 foot portion of the Oxford's mainplane floating in the waves. It was hauled on board. The skipper, Thomas Williamson, signalled his base at Brightlingsea to report the find, but not before tearing off a portion of the yellow fabric from the underside of the wing to keep as a souvenir. The remnant, which bore the black number 3 and part of the number 5, was eventually shown to Pauline Gower who identified it as part of Amy's plane. By a rare twist of fate, Thomas Williamson was a native of Hull and actually knew Amy and her family quite well, although at the time the wreckage was found he had no idea of the identity of its occupant.

Soon after the skipper of the *Young Jacob* sent his signal to the Nore Officer in Command** a boat was sent out from the naval base to collect the wreckage. One is bound to ask why there was such haste to collect this part of the wing, since the *Young Jacob* would have brought it ashore at the end of its patrol in any case. Was it that it might have exhibited signs of damage from our own gunfire? At least one person was convinced that this was the case. Some fifteen years after the end of hostilities, Police Constable Arundel Tucker, who was stationed at West Hampstead in January 1941, told relatives that rumours were circulating in police quarters at that time that examination of

---

* Unlike today, local time in the UK only varied between BST and DBST (Double British Summer Time) during wartime.

** The offices of NOIC were in the Anchor Hotel near the seafront at Brightlingsea.

parts of Amy's aircraft had revealed damage by anti-aircraft fire.[56]

The suspicion that the wreckage might have constituted incriminating evidence is strengthened by the knowledge that neither Crocker nor the ATA were ever allowed to examine the wing, or the part of the Oxford's tail-fin which was washed ashore at Shoeburyness a few days later. When one compares this with the fact that parts of the Messerschmitt Me 110 used by Rudolf Hess on his abortive peace mission to Britain in 1941 are exhibited today in a leading air museum, one cannot but ask why the pieces of wreckage of the aeroplane flown by a national heroine were never deemed important enough to keep. They were quietly disposed of and we are left with nothing more than the theory of a smoking gun.

In 1949 Amy's youngest sister, Betty Falconar Stewart tried to unravel the mystery of her sister's death. She wrote to Lieutenant J.G. Waring, RNR, a native of Hull, who had taken over command of the *Haslemere* in 1943. He told her that he had found 'the greatest difficulty in getting information regarding the incident – nobody was inclined to talk about it'. A few weeks later she wrote to W.D. Kemp, the person in charge of ATA records, and was told, 'Only the Navy had direct evidence of what happened on that day, and, as we found, the Navy are not called "silent" for nothing.'[57]

After the war there was no official court of enquiry into Amy's death, nor was there any attempt to salvage the Airspeed Oxford, even though it sank in quite shallow water on a sandy bed. Its precise position is known and sixty years on, in spite of it being largely of wooden construction, the main portion of the aircraft, including the cockpit and the two Armstrong Siddeley Cheetah engines would still remain reasonably intact. At least three promises have been made by various groups in recent years to investigate the possibility of raising the wreckage, but none has so far materialised.

Alison King glanced up from her desk as Pauline Gower suddenly entered the Operations Room of the Women's Ferry Pool at Hatfield, on that cold and foggy Sunday afternoon. Looking unusually pale, she asked 'Have we had any news of Amy today?'

'No, why do you ask? She's probably forgotten to phone us from Kidlington,' came the slightly nervous reply.

Pauline put a cigarette to her lips. Alison noticed the slight tremor of her boss's fingers as she took it from its silver case. Taking a deep inhalation, she turned to one side and blew a stream of smoke above her head before saying, 'I've just had a phone call from White Waltham aerodrome telling me that two bags have just been fished out of the Thames estuary, and both have "A. Johnson" marked on them.'

'Oh, I wouldn't place too much importance upon that. It's quite a common name, "Johnson".'

'Yes, but they were retrieved from a crashed Oxford.'

'Amy may have put down at Speke or Ternhill in this bad weather. Philippa* had to turn back this morning, didn't she?' Alison could feel her mouth going dry even as she spoke these unconvincing words. She barely managed to keep calm.

Pauline turned to go back into her office and said, 'White Waltham are sending out a signal to all the aerodromes to see if she has landed at any of them. We shall have to wait for their replies until tomorrow, before we go with the news to her family. We've got to be sure of our facts.'

'Shall I ring Prestwick and Kidlington to see what they have to report?'

'Yes, ring around, but don't give anything away – just enquire, that's all.'

Prestwick confirmed that Amy had left there rather late in the day on the Saturday, saying she might put down at Squires Gate if the weather was not too good. Squires Gate reported that she had left that morning at 11.49. Kidlington said that she had not arrived there so far. When asked if there was anything amiss, Alison simply responded, 'No, nothing wrong, just a routine call, that's all.'

Not long after 5 p.m. the phone rang on Alison's desk. She lifted the receiver and a voice asked, 'May I have a word with Miss Johnson, please?' The young Operations Officer was far too shrewd to give anything away; she realised that the caller was a newspaper reporter seeking to confirm a rumour that was already circulating in Fleet Street. 'Sorry, she's not in the office at the moment,' she said. 'Perhaps you could try us tomorrow.'

As soon as Alison reported the phone call to her, Pauline knew that she now had no option but to break the news to Amy's family. If she did not tell them, then the newspapers might, and it would be better coming from her. She sat for a moment deep in thought, reflecting on how Amy had once voiced her feelings about flying on a Sunday. She had told colleagues that it was not a good day for her, since most of her mishaps had occurred on a Sunday. It is interesting to note that the take-off for her Australian flight in 1930 was deliberately delayed until the Monday, as were her successful Cape flights in 1932 and 1936. Pauline went over in her mind exactly how she

---

* First Officer Philippa Bennett had taken off from Hatfield in an Airspeed Oxford during the morning of 5 January 1941 bound for a delivery to Scotland, but at 800 feet, just below the cloud ceiling, ice crystals began to form on her windscreen and she decided to return not long after leaving her base.

was going to tell Amy's parents. She knew of Irene's tragic death, and now there was Amy's – it was not going to be easy for them.

There was a deathly silence in the office throughout that evening. Pauline had already decided to cancel the party on the Monday to commemorate the first anniversary of the Women's Pool. Alison could remember someone remarking that Amy was always meticulous about phoning in when she had put down somewhere unexpectedly. Alison found herself replying, 'Yes, she was always consistent in doing that.' Already, she was being spoken of in the past tense.[58]

Three days after receiving the news of Amy's death, her parents travelled with Betty to stay at the home of John and Alice Hofer in Wooburn Green. The following day they visited the Women's Pool at Hatfield to collect Amy's personal belongings and to find out as much as they could about the circumstances surrounding their daughter's death. Their grief was compounded by the reports that were circulating in the press, to the effect that Amy was carrying a passenger on some unauthorised mission. They were reassured by Pauline Gower that these were no more than unfounded rumours.

They visited William Crocker's office in Gracechurch Street, where they were able to read Amy's will, and to discuss the fact that it could not be implemented until her body was recovered. Little did they realise that almost another three years were to elapse before probate could be established.

On the following Sunday, exactly a week after Amy's death, Will recorded in his diary that 'it was a heavy time for us all between 11.45 a.m. and 3.30 p.m.'. As the clock ticked towards 3.30 Will, Ciss, John, Alice and a friend of the Hofers', Mrs McClaren, stood together in the lounge of Northfield to toast the memory of Amy. As the five glasses were raised, the tears began to fall as a silent tribute was made. The empty glasses were then taken away to be washed and dried. They were then smashed.

The butterfly was gone.

Chapter Twenty-One

EPILOGUE

The era of the long-distance record-breaking pilots and the media frenzy that accompanied them gradually came to an end as the 1930s closed. We look back now on the old black and white cinema newsreels, showing the ecstatic crowds surging towards Lindbergh at Le Bourget or Amy at Croydon with a nostalgia that reminds us of a more innocent age. These were the last of the old frontiersmen and women. They earned their fame and fortune as lone individuals by dint of their own efforts; from now on it would be done by team work. They caught the imagination of the public in a way that has seldom been repeated, and those women who competed in what had hitherto been the sole province of men were especially adulated.

Few would dispute that amongst that pantheon of goddesses from the golden age of aviation, the names that rose above all others were those of Amelia Earhart, Amy Johnson and Jean Batten. To distinguish between their achievements is beyond the scope of this book, but we are left asking ourselves, what made them tick and what they accomplished. Were they merely extrovert women with an eye to the chance for quick fame and fortune? Maybe, but as they each possessed extraordinary courage and a determination to succeed, something which drew the admiration of the ordinary man and woman in the street. Above all, they had the vision to believe that their dreams could be accomplished.

Whilst the immediate aim of all three women must have been self-fulfilment and adventure, they were, by virtue of those aims, effective in promoting 'air-mindedness' and the safety of air travel in an age when aviation as a means of transport was in still in its infancy. Moreover, both Amy and Amelia had the foresight to recognise that the North Atlantic route would eventually become the busiest in the world.

These women also made great strides for the equality of the sexes, whether it was their intention to do so or not. They broke into what was predominantly a man's world and showed that they could compete on equal terms, as witness the way Amy overturned her husband's Cape record in 1932, and that of Tommy Rose in 1936. The American flyer was more overtly feminist, as can be seen by the pre-nuptial agreement she drew up only minutes before her marriage to George Putnam; she also, much like Amy, had the ability to foresee women becoming involved as pilots in military aviation many years before it happened. The contributions of Jean Batten and Amy were on a slightly more muted scale, but equally as effective. They inspired women to change their mind-set, to reach out and to broaden their horizons.

Amy Johnson had the distinction of being the first of the three women to be recognised internationally, when she made her epic solo flight to Australia in 1930. She became as well known throughout the British Empire as she was in the UK. Although Amelia Earhart became the first woman passenger to fly the North Atlantic, she was, as she later admitted, nothing more than 'a sack of potatoes' on the *Friendship* when it was piloted by Bill Stultz and Louis Gordon in 1928. So conscious was she of the fact that her participation in that flight was nothing more than a publicity stunt, one carefully engineered by the man who was to become her husband, that she was only able to throw off the stigma of it when she herself became the first woman to solo the North Atlantic in 1932. Ironically, she never received the acclaim in London for that outstanding solo flight that Amy received in Manhattan after the Mollisons' *Seafarer* crossing in 1933, when they were given the Freedom of the City of New York.

The plucky Jean Batten did not emerge on the international scene until 1934 when, after two failed attempts, she knocked four days off Amy's time from Lympne to Darwin. She then went on to complete four more major flights, including a South Atlantic solo crossing in November 1935, and thus became the first woman to fly from England to South America.

Although Amy did not have the solo transoceanic record flights to her credit that her rivals had, she did risk her neck as co-pilot with Jim Mollison on their mammoth thirty-nine-hour non-stop crossing of the North Atlantic – the first ever from the UK to the USA. Furthermore, who is to say that a forced landing in the ocean is any more perilous than one in the Sahara desert, or descending into the heavily forested and uncharted areas of West Africa.

When it came to endurance on a record-making or record-breaking flight, one could liken Amelia Earhart to a sprinter, whilst the other two women were more like long-distance runners, with more staying power. The well-known Hollywood flyer, Paul Mantz, who flew with Amelia on the first stage of her flight to Honolulu, just before she made her initial attempt at circumnavigating the globe, felt that her physical stamina was suspect, as did her biographer, Mary Lovell. One wonders how she would have coped with the strain of the nineteen-day flight to Australia that Amy undertook in an open-cockpit biplane, as well as the added stress of servicing her own machine at the end of most days' flying.

The emotional make-up of all three women was, as one might expect, quite different. Amy was a romantic with a warm, outgoing personality, a woman who could fall in love easily. She tended to wear her heart on her sleeve. Those who knew her in her early days at Stag Lane remembered how she enjoyed playing football with the men in the hangar and how she bantered and flirted outrageously with them, whereas her two rivals were of a different ilk. Some said of Amelia Earhart that she had a 'dignity and poise' or, as some would put it, an 'aloofness', whilst another, her erstwhile friend Dorothy Putnam, once described her as 'cold-blooded'. But perhaps, in view of the fact that she married Dorothy's ex-husband, we need to treat this opinion with a degree of caution.

With her boyish grin and self-effacing manner, Amelia Earhart had a charm which quickly endeared her to an adoring American public. However, her antipodean rival, Jean Batten, never found it easy to gain acceptance either in her own country, or in the UK. When her biographer, Ian Mackersey, submitted carefully researched material on his subject's background to two independent psychologists, they each found themselves unravelling a very complicated personality indeed. After studying a wealth of documents, ranging from her narratives and personal letters to press-cuttings and transcripts of her filmed and tape-recorded interviews, they came to almost identical conclusions. They found her to be 'classically androgynous'; in other words to have the sexually attractive body of a woman, but the inner drive of a man.[59] She was a completely self-absorbed person, a woman who saw herself as having a destiny that had to be fulfilled at all costs, so much so that she could display a rudeness and an arrogance which put many people against her.

By some strange twist of fate all three women left a legacy of mystery at the end of their lives. The bodies of Amelia and Amy were never found, and for many years neither was that of Jean Batten. After the war Jean moved with her mother successively to Jamaica, Spain, and then to Tenerife, where her elderly parent died. Friendless, lonely and no longer an international celebrity, she moved into a small, seedy hotel in the southern suburbs of Palma, Majorca, where for almost five years she seemed to disappear completely without trace. It was discovered that she had died from an untreated infection in her leg caused by a dog bite. Through some bureaucratic mix-up by the Spanish authorities, her death on 22 November 1982 was not reported to the British Consulate on the island, and her body was interred in a communal grave along with more than fifty others. The only reference to her that could be found amongst cemetery records when enquiries were made, was the simple entry: 'Jean Gardner Batten, burial number 146.' All that remains in her memory today are two simple bronze plaques, one attached to the wall above the communal grave in Palma; and another in the entrance hall of the apartment block in the Avenida del Generalissimo Franco, Puerto de la Cruz, where she once lived.

Will and Ciss Johnson remained with the Hofer family at Wooburn Green until Tuesday, 14 January 1941, when they attended a memorial service at St Martins in the Fields in honour of their daughter. Lord Beaverbrook, Minister for Aircraft Production, and Sir Archibald Sinclair, the Air Minister, were represented, together with Caroline Haslett, President of the Women's Engineering Society, and many of Amy's colleagues from the various ferry pools of the Air Transport Auxiliary. Tributes poured in from around the world and her obituary appeared in every section of the national press, mourning the loss of a national heroine. In order to keep her memory alive, an appeal was launched by the WES for the founding of a scholarship in aviation for women. This is overseen today by the Amy Johnson Memorial Trust.

It seemed that the Johnson family were destined never to be very far from human tragedy, and Amy's parents received yet another hammer blow when their son-in-law, Squadron Leader Ronald Falconar Stewart, DFC, was killed in action three years later. The twenty-three-year-old RAF pilot and his Lancaster crew of five perished on the night of 1/2 January 1944 whilst acting as one of the pathfinder force marking the target and leading an attack of 421 bombers on Berlin. Like many of the other twenty-seven Lancasters that failed to return that night, they fell victim to the German night fighters. Betty, already scarred by the suicide of her older sister, Irene, and the untimely death of Amy, was now devastated by the loss of her husband, and it left her suffering from a recurring clinical depression. Sadly, she took her own life in 1973. Mercifully, her mother and father were spared this further tragedy, having themselves passed away by that time.

One might well ask:'What happened to the two men in Amy's life?'

Jim Mollison married for the third time in September 1949 after meeting the attractive and wealthy widow, Mary Kamphuis, former wife of a Dutch Air Force colonel. The marriage foundered six years later, mainly due to Jim's heavy drinking, and they agreed to separate. Mary made a generous settlement on her husband, with which he was able to purchase the Carisbrooke Hotel in Surbiton, Surrey, giving him, as she thought, some purpose in life. Although his wartime service with the ATA had enabled him to exercise a certain amount of self-control over his drinking habits, he now had little left to restrain him. He was a misfit in the postwar world, without the celebrity status he once enjoyed. He found it difficult to adapt to a quieter life now that his name was no longer recognised by a younger generation.

By 1959 he was diagnosed as suffering from alcoholic epilepsy, one of the severest forms of alcoholism, and his health deteriorated rapidly. He suffered a severe blackout in the October and was rushed into the Priory Hospital at Roehampton. Life in the fast lane had taken its toll. He remained in a coma for just over a week and died without regaining consciousness. So ended the life of the man who was, arguably, the greatest of all the long-distance solo aviators.

And what of Hans? It has been said that one's first love is often the strongest, and in the case of the intense relationship that existed between Amy and Hans for six years, there is good reason to believe that this was so. The emotional wounds he inflicted on her were slow to heal and on the rare occasions when she did bare her soul on the subject, she showed how deeply she was hurt. When Sir James Martin visited her lodgings not long after she moved to Roe Green to be nearer to her flying tuition at Stag Lane, he noticed a rather beautiful painting of Swiss mountain scenery on the wall of her bed-sit. When he asked her if she had ever visited that country, she became quite tearful and told him how she had been given the picture as a wedding present in the belief that she and her Swiss boyfriend would marry one day. She then told him how her lover had thrown her over to marry another girl, and how she had contemplated suicide at that time.

For over forty-five years Amy's lover kept the several hundred letters that she had written to him, and only released them into the public domain shortly before he died from a degenerative illness in 1969. The one thing that Amy was never to know was that the marriage of the man whom she regarded as 'not the marrying kind', produced a child just six months after his wedding day.

With families who have lost a loved one and where the body of that person is never recovered, there is always an incompleteness in their grieving. This feeling was to haunt both Amy's parents and her sisters. Old wounds were reopened for them when the skeleton of a woman was uncovered on the foreshore near Herne Bay in February 1961. A rumour quickly sprang up in the press that the remains were those of Amy, but when her dental records were consulted the rumour proved to be false. A further heartache might well have arisen if they ever heard the apocryphal and ghoulish story which circulated amongst the fishermen of the Essex coastal waters. It was said that the body of a woman in military uniform had been recovered from the sea, caught up in a trawler's net. Not wishing to have to make an appearance at an inquest, the skipper ordered it be dropped back overboard.

Ciss and Will Johnson lived next door to their daughter Mollie and her family in the early postwar years, before retiring to their native Yorkshire. Mollie's two daughters still treasure happy and fond memories of their grandparents during their time in Blackpool, and of their grandpa in particular. He was a keen amateur ciné photographer throughout his lifetime, and was rarely seen without his beloved 8 mm camera on special family occasions. He loved to potter around in a room over the garage, a favourite spot of his, which was known to the family as Grandpa's den. It was where he would show his old movies to the two young girls. One of their lasting impressions is of wreaths of cigar smoke drifting through the beam of the projector as he entertained them with one of his 'funnies'; and of the squeals of laughter whenever a spool of film would occasionally spill over onto the floor.

Ciss Johnson died in 1958 leaving her husband to live on for a further five years at his home in Woodmansey, near Beverley. The old man rarely spoke of his two eldest daughters; it was probably far too painful for him to do so, but no doubt, he lived on past memories. He was a meticulous diarist and he would record both Irene and Amy's birthdays on the appropriate day, as if they were still alive, as indeed they were in his mind. Those who die young will always stay young.

If by definition heroism means to be distinguished by exceptional courage and extraordinary strength, then Amy Johnson qualified on both counts. Not only was she an outstanding woman of the 20th century, vying with those who made great strides for the emancipation of women, but she is arguably one of the last great heroines.

Although she may have failed to gain in her private life what she longed for most – to be truly loved by the two men in her life, what cannot be denied is the strong bond of love that existed between father and daughter. To him, she was immortal.

Perhaps the most poignant entry in Will Johnson's diary reads: 'Last night I dreamt that Amy came and kissed me. It was very real. I believe it *was* real.'

BIBLIOGRAPHY

Barnato Walker, Diana, *Spreading My Wings*, Patrick Stephens Ltd., 1994.
Balfour, Christopher, *Spithead Express*, Magna Press, 1999.
Bergel, Hugh, *Fly & Deliver*, Airlife, 1982.
Bramson, Alan, *Master Airman*, Airlife, 1985.
Chapman, Putnam, Sally, & Mansfield, Stephanie, *Whistled Like a Bird*, Warner Books, 1997.
Cheesman, E. C., *Brief Glory*, ATA Association, 1995.
Cluett, D., *Croydon Airport – The Australian Connection*, Sutton Libraries, 1988.
Cluett, D., Nash, J., & Learmonth, R., *Croydon Airport – The Great Days, 1928–1939*, Sutton Libraries, 1980.
Cobham, Sir Alan, *A Time to Fly*, Shepheard-Walwyn, 1986.
Cochran, Jacqueline, & Brinley, M. B., *Jackie Cochran*, Bantam Press, 1987.
Courtenay, William, *Airman Friday*, Hutchinson, 1937.
Crocker, Sir William Charles, *Far from Humdrum*, Hutchinson, 1967.
Cross, Wilbur, *Zeppelins of WW1*, I. B. Tauris – London & NY, 1991.
Curry, Robert, *Last Complete Performance*, Hutton Press/Hull College of Further Education, 1992.
Curtis, Lettice, *The Forgotten Pilots*, Nelson & Saunders Ltd., 1982.
Fahie, Michael, *A Harvest of Memories*.
Finch, Robert, *Amy Johnson: Global Adventurer*, Hull College of Further Education, 1989.
Fitzpatrick, Eva, *Cross Country*, Hothersall & Travers, 1989.
Fox, James, *White Mischief*, Penguin Books, 1988.
Foynes, J. P., *Battle of the East Coast*, Self-published, 1984.
Frater, Alexander, *Beyond the Blue Horizon*, Penguin Books, 1987.
Gibbs, Roger, *Cresta Run 1885–1985*, Henry Melland, London, 1986.
Gillett, Edward, & MacMahon, Kenneth, A., *A History of Hull*, Hull University Press, 1989.

Grey, Elizabeth, *Winged Victory*, Constable, 1966.

Halpenny, Bruce, Barrymore, *Action Stations No: 8*, Patrick Stephens Ltd., 1984.

Henshaw, Alex, *The Flight of the Mew Gull*, John Murray Ltd., 1980.

Hull, Edward, *Take up Slack*, Woodfield Publishing, 2000.

Johnson, Amy, *Myself when Young*, Frederick Muller Ltd., 1938.

Johnson, Amy, *Skyroads of the World*, W & R Chambers, 1939.

Johnstone, 'Sandy', CB, DFC, *Spitfire into War*, William Kimber & Co Ltd., 1986.

King, Alison, *Golden Wings*, C.A. Pearson Ltd., 1956.

Lomax, Judy, *Women of the Air*, John Murray, 1986.

Lovell, Mary, S., *Straight on Till Morning*, Arena, 1988.

Lovell, Mary, S., *The Sound of Wings*, Hutchinson, 1989.

Luff, David, *Mollison - The Flying Scotsman*, Lidun Publishing, 1993.

Mackersey, Ian, *Jean Batten: The Garbo of the Skies*, Warner Books, 1992.

McKee, Alexander, *Coal Scuttle Brigade*, Souvenir Press, 1957.

McKee, Alexander, *Into the Blue*, Souvenir Press, 1981.

Mollison, James, Allan, *Death Cometh soon or Late*, Hutchinson, 1932.

Mollison, James, Allan, *Playboy of the Air*, Michael Joseph, 1937.

Oakes, Claudia, M., *U.S. Women in Aviation, 1930-39*, S.I.P., 1991.

Parry, Simon, W., *Intruders Over Britain*, Air Research Publications, 1987.

Robson, Graham, *The Monte Carlo Rally*, B.T. Batsford, 1989.

Rose, Norman, *Vansittart - Study of a Diplomat*, Heineman, 1978.

Scott, C.W.A., *Scott's Book*, Hodder & Stoughton, 1934.

Sharman, Sarah, *Sir James Martin*, Patrick Stephens Ltd., 1996.

Sharp, Martin, C., *D H - A History of de Havilland*, Airlife, 1982.

Smith Babington, Constance, *Amy Johnson*, William Collins, Sons & Co, 1967.

Snell, Gordon, *Amy Johnson - Queen of the Air*, Hodder & Stoughton, 1980.

Sumner, Ian & Margaret, *Bridlington*, Alan Sutton Publishing, 1995.

Thacker, Tony, *'32 Ford - The Deuce*, Osprey, 1984.

Wadsworth, Michael, *They Led the Way*, Highgate Publications (Beverley) Ltd., 1992.

Some articles consulted:

Crawley, Paul, *Amy Johnson and the German agent*, 1997; Gill, Alec, *What's Amy Johnson got to do with Hessle Road?* 1994; Hicks, John, *From Boulevard to Kingston (1895-1990 The Story of our School)*, 1990; Johnson, Amy, *A Day as a Ground Engineer*, The Woman Engineer, 1930; *A Day's work in the ATA*, The Woman Engineer, 1940; *What Next?* Aug 1938; *The Life of an Airwoman*, Lilliput, 1939; King, Alison, *Amy's Last Flight*, Aeroplane Monthly, Parts I & II, Nov & Dec 1980; Mollison, Mr & Mrs, *The Value of Record-breaking Flights - a debate*, The Woman Engineer, Jan 1935; Murtagh, Michael, *Why can't we have radio?* Aeroplane Monthly, Aug 1974; Nesbit, Roy, *What Did Happen to Amy Johnson?* Aeroplane Monthly, Parts I & II, Jan & Feb 1988; Whitaker, George, *Cardiff's Municipal Aerodrome*; Wynn, H., *What Happened to Amy?*

Acknowledgements

I am deeply indebted to Amy Johnson's surviving relatives, Judith Verkerk and her sister Susan Crook, for copyright permission to quote from the aviatrix's private letters, books and articles. Without their generous support and co-operation the book could never have been written. Their one desire throughout the whole project of compiling this official biography has been to see their late aunt's memory preserved and portrayed in a factual and balanced way, without seeking to promote a hagiography or to contemporise her history in any way as some might be tempted to do. I trust that in meeting their expectations, I will also meet with the approval not only of those concerned more readily with aviation history, but with the general reader who wants to know more about the inner life of this outstanding woman.

I wish to give a special thanks to my old friend and colleague, Peter Little, an *aficionado* of all things concerning the aviatrix, and one possessing an unrivalled private collection of Amy Johnson memorabilia. He has given unstinting support and been a tremendous help especially during my research period. One other person who deserves a special mention is Paul Crawley, who has been most helpful in scouring archives concerning Amy's time in Cardiff; as well as supplying me with copies of the depositions given during the Probate Court of Inquiry in December 1943.

Another to whom I wish to record my appreciation is Eva Fitzpatrick, for her kindness in allowing me to quote lengthy extracts from her excellent book, *Cross Country*, and for the photograph of her late father, Herbert G. Travers. As an eye-witness to events during the time Amy was learning to fly, she has contributed a whole new aspect of the feuding that went on at Stag Lane when it was touch and go on whether Lord Wakefield would back the Australia flight in 1930.

My thanks also to the late Mollie Jermey, sole beneficiary of the James Allan Mollison estate for permission to quote from his autobiography, *Playboy of the Air*.

Space does not permit further detail of the help and encouragement I have received from so many, but I now list those who have contributed so generously in various ways:

Colin Ashford, Christopher Balfour, Ben Barker, Douglas Barltrop, Victor & Margery Bitter, USA; Russell Brown, the Fylde aviation historian; Grizelda Chlebowska, sister of the late Ronald George Falconar Stewart, DFC; Eddie Clipperton, Gerald Edwards, Harry Fidler, Jack Fuller of the Croydon Airport Society; William A. Harrison, Alex Henshaw, The late John D. Hicks, Jo Hodder of the Society of Authors; Brian Goulden, Revd. James D. Hargreave, Muriel Hollington, Tony Hill, Edward Hull of the London Gliding Club; Vera Johnson, Barbara Morgan-Jones, Christopher Ketchell of the Amy Johnson Appreciation Society; Simon Kidston, Line Larin of the Ritz-Carlton Hotel, Montreal; Colin Latham, May Maple, Andrew Marr, Stuart McKay of the de Havilland Moth Club; William Mollison, Tom Mitchell, Geoffrey Morris of the C.A.S.; J. A. Peck, Headmaster of Kingston High School, Hull; Greg Piddocke, Marjorie Pocock, Royston Powell (for his help in wartime naval matters); Harry & Pat Purkis of the A.J.A. Society; Norman Rigler, Norman Staveley, Patricia Stewart (for research material on Gwyneth Roulston and Winifred Irving); Ian & Margaret Sumner, Muriel E. Tucker, Stuart V. Tucker, Sheila Waddell, Diana Barnato Walker, Melanie Whittaker and Edmund Wynne.

Photographs form an important part of any biography and I wish to record my special thanks to the following contributors:

To Judith Verkerk and Susan Crook for allowing me to use family photographs from the Amy Johnson Collection; Robert Barnard, D. Holyland of Martin-Baker Ltd.; Roger T. Jackson, Dr. D. J. Marchant, Ron Neudegg, Brian Riddle of the Royal Aeronautical Society; Richard Riding, Mrs Shann Stokes (née Puckle) and Wing Cdr. Eric Viles of the Air Transport Auxiliary Association.

Museums & Libraries
John Sanford of the Wright State University, Dayton, Ohio; Peter J. V. Elliott of the Royal Air Force Museum, Hendon; The Canadian Embassy; Ian Williams of the Ansdell Library, Lytham St. Annes; Bill Torrens of the Local Studies Library, Aylesbury; J. Marsh of the Reference Library, Bournemouth; Bridlington

Central Library; Miss M. Jones of Eastbourne Central Library; Zoe Gray of Gravesend Central Library; Sylvia McKean of Herne Bay Central Library; Mrs Jo Edge and David Smith of the Local Studies Library, Hull; and Philip Shuttleworth of St.Annes Library, Lytham St.Annes.

Lastly, and by no means least, I wish to express my gratitude to my wife Vera for her love, support and encouragement throughout the period of research and writing.

Every effort has been made to trace all sources of extracts and illustrations used throughout this book and apologies are given if there has been any infringement of rights. The publishers will be glad to make good in future editions any error or omission brought to them.

Source Notes

1. *Garbo of the Skies* by Ian Mackersey.
2. *Journal of The Society of Engineers*, Vol. XXI, No: 4, Oct–Dec 1930.
3. *A Time to Fly* by Sir Alan Cobham.
4. *Journal of The Society of Engineers*, ibid.
5. Ibid.
6. Ibid.
7. Îbid.
8. *Mollison - The Flying Scotsman* by David Luff.
9. *Amy Johnson* by Constance Babington Smith.
10. *Airman Friday* by William Courtenay.
11. *Croydon Airport Society*; Newsletter 66/96.
12. *Scott's Book* by C.W.A. Scott.
13. *Croydon Airport Society*; Newsletter 41/90.
14. *Scott's Book*.
15. *Skyroads of the World* by Amy Johnson.
16. *Cresta Run 1885-1985*: Bob Ennis.
17. *Airman Friday*.
18. *Playboy of the Air* by James Allan Mollison.
19. Correspondence with Max MacLeod, September 1990.
20. *September Champions - The Story of America's Air Racing Pioneers* by Robert Hull.
21. *Playboy of the Air*.
22. Ibid.
23. *Sky Fever* by Sir Geoffrey de Havilland.
24. Correspondence with Jeffrey Quill, January 1990.
25. *Scott's Book*.
26. Correspondence with Jehangir R. D. Tata, October 1989.
27. *White Mischief* by James Fox.

28. Daniel O'Doheny via Line Larin of Ritz-Carlton Hotel, Montreal, February 2000.
29. *West Lancashire Evening Gazette*, 12 January 1988 and telephone conversation with wife of the late W. H. Davies.
30. *Lilliput*, Pocket Omnibus Edition, December 1939.
31. *32 Ford – The Deuce* by Tony Thacker.
32. *Lilliput*, ibid.
33. Ibid.
34. *Spithead Express* by Christopher Balfour.
35. Correspondence and phone discussion with G. Edwards, May 1999.
36. *South Wales Echo & Evening Express*, 3 October 1939.
37. *Vansittart – Study of a Diplomat* by Norman Rose.
38. *The Aeroplane*, circa 1940.
39. *Golden Wings* by Alison King.
40. *Brief Glory* by E. C. Cheesman.
41. Ibid.
42. *Spitfire into War* by Sandy Johnstone.
43. Correspondence from Greg Piddocke via Wing Cdr Eric Viles, MBE, November 1997.
44. Correspondence with Dennis Turberfield, February 1989.
45. Alison King, *Aeroplane Monthly*, November 1980.
46. *Southend Evening Echo*, 15 February 1988.
47. Roy C. Nesbit, *Aeroplane Monthly*, February 1988.
48. *Spreading My Wings* by Diana Barnato Walker, MBE.
49. *Aeroplane Monthly*, November 1980.
50. Interview with Ernest L. Hannam, June 1988.
51. Roy C. Nesbit, *Aeroplane Monthly*, February 1988.
52. Correspondence and phone discussions with Thomas Mitchell during the period March–December 2000.
53. Phone discussions with Melanie Whitaker of the *Kent & Sussex Courier* during the same period as above.
54. Phone discussions with Richard Powell in September 1991 and July 1997.
55. Article by Leslie Hunt, MBE, 11 January 1991.
56. Phone discussion with Stuart V. Tucker, September 1991.
57. Betty Falconar Stewart's correspondence with Lt. J. G. Waring, RNR, 14 November 1949; and with W. D. Kemp, i/c ATA Records, 25 November 1949.
58. *Golden Wings*.
59. *Garbo of the Skies*.

INDEX

Pages in heavy type represent photographs, illustrations, captions and footnote references.